5.5.3 Single Isomer Negatively Charged Cyclodextrins	S-68
	5.5.4 Positively Charged Cyclodextrins	S-68
	5.5.5 Mixed Selector Systems	S-69
	5.5.6 Dual Cyclodextrin Selector Systems	S-71
5.6	Analyte Classes	S-72
	5.6.1 Metabolism Studies	S-72
	5.6.2 Process Analysis	S-72
	5.6.3 Forensic Analysis	S-72
	5.6.4 Carbohydrates and Peptides	S-73
	5.6.5 Alkaloids	S-73
	5.6.6 Herbicides.	S-73
5.7	Analyte Structure.	S-74
	5.7.1 Stereogenic Nitrogen Atoms	S-74
	5.7.2 Stereogenic Sulphur Atoms	S-74
	5.7.3 Hindered Rotation	S-75
5.8	Batch and Source Variation	S-75
5.9	Conclusion	S-75
6	**Other Chiral Selectors**	S-78
6.1	Introduction	S-78
6.2	Ligand Exchange Selectors.	S-78
6.3	Chiral Surfactants	S-79
	6.3.1 Synthetic Chiral Surfactants	S-79
	6.3.2 Naturally Occurring Chiral Surfactants	S-79
6.4	Alkaloids	S-80
6.5	Crown Ethers	S-81
6.6	Natural Macromolecules	S-82
	6.6.1 Proteins.	S-82
	6.6.2 Macrocyclic Antibiotics	S-83
	6.6.3 Polysaccharides	S-86
6.7	Non-Aqueous CE	S-86
6.8	Derivatisation	S-87
6.9	Capillary Electrochromatography	S-88
6.10	Conclusion	S-91
	Index	S-93
	Information to previous publications in this series. . .	S-6

The study of enantiomers is important because of the potential for differences in the changes they can produce in living systems. Analytical methods which separate enantiomers are tools which can make a major contribution to our understanding of the differences in their behaviour.

In the first chapter the author describes CE as an important additional tool for scientists interested in studying enantiomers and ensuring their quality and safety. After that he gives a short overview about the theory of the technique and the basic separation principles. The key to the separation of enantiomers in CE therefore lies in the introduction of additional agents which induce differences in their effective charges, sizes, or shapes. To understand the separation and resolution of enantiomers using CE there follows a discussion of the various physical and mathematical models and the underlying physical processes and interactions.

The use of a systematic approach to screening chiral selectors and optimising experimental conditions is also beneficial. In the last chapters the author therefore describes (i) cyclodextrins as a popular group of chiral selectors for the separation of enantiomers by CE and (ii) some of the selectors other than cyclodextrins which are used as buffer additives to separate enantiomers in free solution CE.

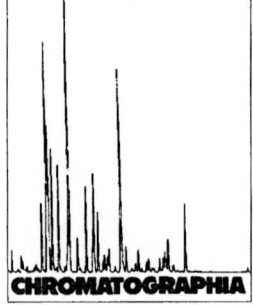

Vol. 54, Supplement 2001

Chromatographia was founded by R. E. Kaiser in 1968.

Publisher

Chromatographia is published by Friedr. Vieweg & Sohn Verlagsgesellschaft mbH, P.O. Box 1546, D-65173 Wiesbaden, Federal Republic of Germany, Tel. +49(0)611 7878 380(–381); Telefax +49(0)611 7878 439
Editorial office e-mail: CHROMATOGRAPHIA@bertelsmann.de; webpage: http:\\www.chromatographia.de
For more information regarding Vieweg's program for books and journals see our homepage: http://www.vieweg.de

Advertising Representatives

Inquiries concerning advertising should be addressed to the publisher's address above;
Tel. +49(0)611 7878 153, Fax –430.
Inquiries in USA: Trade Media International, 424 Madison Avenue, New York, NY 10017, USA; Tel. (212)421–1229.
Inquiries in the UK: Elsevier Science Ltd., The Boulevard, Langford Lane, Kidlington, Oxford, OX5 1GB, UK.

Distributors

Friedr. Vieweg & Sohn, P.O. Box 1546, D-65173 Wiesbaden, Germany;
Tel. +49(0)611 7878 324; Telefax +49(0)611 7878 423.
Elsevier Science Ltd., The Boulevard, Langford Lane, Kidlington, Oxford, OX5 1GB, UK.

Distributions in the USA

Chromatographia (USPS No. 374810) is distributed by German Language Publications, Inc., 153 South Dean Street, Englewood, NJ 07631. Second class postage is paid at Englewood, NJ 07631.
Postmaster: send address changes to Chromatographia, German Language Publications, Inc., 153 South Dean Street, Englewood, NJ 07631.

Subscriptions

Chromatographia is published monthly. Up to three volumes may be published per year.

Vols. 53, 54, Supplements (2001) (approx. 2440 pp.)	DM 2.364,–	US $ 1,243.00	öS 17.257,–	sFr 2.103,–	€ 1.209,–
Single copy	DM 224,–	US $ 118.00	öS 1.635,–	sFr 199,–	€ 115,–

For individual subscribers who will certify that Chromatographia is for their personal use only (to be ordered directly from the publisher):

Vols. 53, 54, Supplements (2001)	DM 1.344,–	US $ 706.00	öS 9.811,–	sFr 1.196,–	€ 687,–

All prices include postage. Subscriptions are renewed automatically for one year unless notice to terminate the subscription is given three months before the end of the current year.

Submission of Papers

One original and two copies of manuscripts should be sent to Chromatographia, Editorial Office at the same address as the publisher, above. For papers intended for review-type articles, an outline of the proposed article should first be forwarded to the Editorial Office Manager for preliminary discussion, prior to preparation. For general information on the rules concerning style and format of manuscripts please refer to "Instructions to Authors" in every issue.

Softcover reprint of the hardcover 1st edition 2001
Vieweg is a company in the specialist publishing group BertelsmannSpringer

Typesetting: Fotosatz Behrens, Oftersheim.
Printed on acid-free paper.
ISBN 978-3-322-83143-9   ISBN 978-3-322-83141-5 (eBook)
DOI 10.1007/978-3-322-83141-5

Chromatogram on front page: Gas chromatographic separation of gasoline hydro carbons (selected section of chromatogram) with glass capillary column.

CHROMATOGRAPHIA

An International Journal for Rapid Communication in Chromatography, Electrophoresis, and Associated Techniques

Abstracted in Anal. Abstr., ASCA. Biodet. Abstr., Biol. Abstr., Cadscan, Chem. Abstr., Chem. Cit. Ind., C.I.S. Abstr., Current Contents, Deep Sea Res. & Oceanogr. Abstr., Diary Sci. Abstr., Excep. Med., Food Sci. & Techn. Abstr., GeoRef., INIS Atormind. Ind. Sci. Rev., Ind. Vet., Lead Abstr., Mass Spectr. Bull., Nat. Sci. Cit. Ind., Rev. Med. & Vet. Mycol., Sci. Cit. Ind., Sel. Water Res. Abstr., Sugar Ind. Abstr., Vet. Bull., VITIS, Weed Abstr., W.R.C. Inf., Zine Scan

Volume 54, Supplement 2001

Editorial Office

M. Schaub, Manager

Vieweg Publishing
P.O. Box 1546
65173 Wiesbaden, Germany

Tel. +49 (0)611 7878 380, 381
Fax +49 (0)611 7878 439

H. Weinheimer, Publisher

vieweg

CHROMATOGRAPHIA

An International Journal for Rapid Communication in Chromatography, Electrophoresis, and Associated Techniques

Contents Supplement Vol. 54, 2001

Preface S-5

1   Enantiomers and Separation S-7
1.1   Introduction S-7
1.2   Enantiomers and Living Organisms S-8
    1.2.1 Background S-8
    1.2.2 Regulatory Aspects....................... S-9
    1.2.3 Chiral Switches S-10
    1.2.4 Enantiomeric Agrochemicals S-10
1.3   The Separation of Enantiomers................... S-10
    1.3.1 Chromatographic Separation of Enantiomers S-10
1.4   Electrophoretic Separation of Enantiomers S-11
    1.4.1 The Development of Capillary Electrophoresis S-11
1.5   Enantiomer Separations by Capillary
    Electrophoresis............................... S-12
    1.5.1 Ligand Exchange........................ S-12
    1.5.2 Cyclodextrins........................... S-12
    1.5.3 Chiral Surfactants....................... S-13
    1.5.4 Mixed Cyclodextrin and Surfactant Systems .. S-13
1.6   Conclusion S-13

2   The Principles of Separation in CE S-15
2.1   Introduction: Movement and Separation S-15
2.2   Separative and Non-Separative Transport in CE.... S-16
2.3   Electrophoretic Mobility S-16
    2.3.1 Molecular Size.......................... S-16
    2.3.2 Molecular Charge S-17
    2.3.3 Molecular Shape S-17
    2.3.4 Buffer Type and Concentration S-18
2.4   Electroosmotic Mobility........................ S-19
    2.4.1 Background Theory S-19
    2.4.2 Experimental Conditions................... S-20
2.5   CE Instrumentation............................ S-21
    2.5.1 Introduction............................ S-21
    2.5.2 The Power Supply S-21
    2.5.3 The Capillary........................... S-21
    2.5.4 The Injector S-22
    2.5.5 The Detector S-22
    2.5.6 The Data System S-22
2.6   Conclusion S-23

3   Modelling Enantiomer Separation by CE S-24
3.1   The Use of Models............................ S-24
3.2   Background S-24
3.3   The Physical Model........................... S-25
3.4   A Basic Mathematical Model S-26
    3.4.1 Electrophoretic Mobility................... S-26
    3.4.2 Electrophoretic Mobility Difference S-28
    3.4.3 An Analytical Solution.................... S-30

    3.4.4 Interpretation of Previous Data S-31
    3.4.5 Experimental Support S-31
    3.4.6 Resolution S-36
3.5   Extended Theoretical Models S-37
    3.5.1 Different Limiting Mobilities S-37
    3.5.2 Equilibria Involving more than
        one Chiral Selector S-38
    3.5.3 Equilibria Involving other Species S-39
3.6   Conclusion S-40

4   Method Development Strategies. S-42
4.1   The Purpose of the Method S-42
4.2   Factors Controlling Resolution S-43
    4.2.1 Efficiency S-43
    4.2.2 Selectivity S-44
4.3   The Structure of the Analyte S-45
4.4   The Choice of Chiral Selector S-46
    4.4.1 Analyte Variation S-47
    4.4.2 Chiral Selector Variation S-48
    4.4.3 Screening Selector Type and Concentration... S-48
4.5   Other Separation Conditions.................... S-49
    4.5.1 The Analyte S-49
    4.5.2 Choice of Capillary....................... S-49
    4.5.3 Buffer pH S-49
    4.5.4 The Choice of Buffer and Buffer Concentration S-50
    4.5.5 Electrical Field Strength.................. S-53
    4.5.6 Operating Temperature S-53
4.6   Optimisation Approaches S-53
    4.6.1 Univariate Optimisation S-54
    4.6.2 Multivariate Optimisation.................. S-54
    4.6.3 Suggested Approaches S-55
4.7   Validation S-57
4.8   Computer Simulations S-57
4.9   Conclusion S-57

5   The Use of Cyclodextrins as Chiral Selectors S-59
5.1   Introduction.................................. S-59
5.2   Structure and Properties........................ S-59
5.3   Cyclodextrin Production S-60
5.4   Complexation Mechanisms S-61
    5.4.1 Equilibrium Constants..................... S-61
    5.4.2 Evidence for Inclusion S-61
    5.4.3 The Role of the Organic Solvent............. S-62
    5.4.4 Thermodynamics of Inclusion S-62
    5.4.5 The Importance of pH S-63
    5.4.6 Changes in Migration Order S-64
5.5   Cyclodextrin Classes S-65
    5.5.1 Neutral Cyclodextrins S-65
    5.5.2 Negatively Charged Cyclodextrins........... S-66

Preface

It is now over twenty years since the publication of the pioneering investigations by Jorgenson and other workers on modern capillary electrophoresis, and over ten since the introduction of commercial high performance instruments suitable for automated analysis. During this time capillary electrophoresis has been transformed from a curiosity into a routine tool in industrial and academic laboratories throughout the world.

Capillary electrophoresis is now accepted alongside more mature separation techniques such as HPLC, and has established itself as a valuable complimentary approach in a number of areas of importance. One of the areas where capillary electrophoresis has had a significant impact is the subject of this book, the separation of enantiomers. Enantiomer separation is of particular importance in pharmaceutical analysis and capillary electrophoresis is beneficial because of the high efficiencies generated.

In this work I have tried to show both how and why capillary electrophoresis can be beneficially applied to enantiomer separation. I have attempted to cover the basic principles of movement and separation, how these principles can be used to develop theories describing enantiomer separation, and then how the theory can be used to assist the practitioner in the development and optimisation of methods suitable for routine use.

The field of enantiomer separation is obviously large and so any review is inevitably selective. In this book I have not attempted a detailed summary of the complete body of publication, but rather to try and provide the reader with some useful insight.

I would like to express my thanks to friends and colleagues who have assisted in the writing of this book via their support, advice, patience and constructive comments.

Chinley, June 2001   Stephen Wren

CHROMATOGRAPHIA

CE Series

Edited by Kevin D. Altria, Glaxo Wellcome R&D, UK

There are currently a number of general textbooks covering Capillary Electrophoresis where information on a range of applications and techniques can be found. Readers who are interested in a specific area of CE struggle to find truly comprehensive treatments of their areas of interest. The CHROMATOGRAPHIA CE series has been established to allow comprehensive books to be produced covering individual topics. The books are written by well known authors in their specialist application areas and cover CE topics such as DNA analysis, analysis of pharmaceuticals, chiral separations, MECC, carbohydrate analysis, biomedical applications and troubleshooting in CE.

- **Volume 1:** C. Heller (Ed.), Analysis of Nucleic Acids by Capillary Electrophoresis

- **Volume 2:** K. D. Altria, Analysis of Pharmaceuticals by Capillary Electrophoresis

- **Volume 3:** A. Paulus / A. Klockow-Beck, Analysis of Carbohydrates by Capillary Electrophoresis

- **Volume 4:** W. Kok, Capillary Electrophoresis: Instrumentation and Operation

- **Volume 5:** A. B. Chen / W. Nashabeh / T. Wehr (Eds), CE in Biotechnology: Practical Applications for Protein and Peptide Analyses

- **Volume 6:** S. Wren, The Separation of Enantiomers by Capillary Electrophoresis

1 Enantiomers and Separation

1.1 Introduction

> Tyger! Tyger! burning bright
> In the forest of the night,
> What immortal hand or eye
> Could frame thy fearful symmetry?
>
> William Blake (1757–1827)

Symmetry exists in many guises and forms in the natural world and has profound significance to our lives. The concept of symmetry has a broad appeal and so is applied to what we perceive with our senses and right down to the sub atomic level. The Tiger for example has a plane of reflectional symmetry running down the centre of her nose which bisects her body. An additional feature of interest for the chemist is that the symmetry of the Tiger is passed onto her cubs and so conserved through the generations.

Symmetry is also as important a feature of the microscopic world as it is of the macroscopic. One of the areas where symmetry has a profound influence on the microscopic world is the subject of this book, the reflectional symmetry which gives rise to enantiomeric pairs.

Enantiomers are isomeric pairs of molecules which are mirror images of each other that are not superimposable. This form of isomerism arises because molecules are three dimensional objects and so the atoms may be arranged in more ways than would be possible with a planar two dimensional arrangement. An example is the tetrahedral arrangement of atoms or groups around a tetravalent carbon atom. In Figure 1.1 the two possible arrangements of a carbon atom bound to the atoms or groups W, X, Y, and Z are shown.

At this point we note the difficulty of trying to depict arrangements which exist in three dimensions when limited to the use of a two dimensional medium – a flat sheet of paper. Because of the limitations of the written page there is no substitute

for the use of framework or other molecular models as a means of more fully understanding the issues. In order to alleviate the difficulties which arise from depictions of three dimensional objects in two dimensional we commonly resort to the use of conventions to represent molecules. Atomic bonds which should project towards the reader are drawn as solid wedges and bonds which project away from the reader are represented as broken lines.

The two forms in 1.1a and 1.1b can be seen to be mirror images of each other. The two forms are not however identical and cannot be superimposed. For example rotation of the molecule through 180 degrees about the W–C bond would produce 1.1c. The later depiction has the same arrangement of W, C, and X, as Figure 1.1a, but has Y and Z transposed.

The phenomenon of enantiomeric forms is important because the two forms can interact with the physical world in different ways. Of particular interest to us as living beings is the fact that the enantiomeric forms can interact differently with biological systems. It is this fact which has been the main driving force in the huge interest in the selective synthesis and analysis of enantiomers over the last few decades.

The physical interaction which defines and differentiates between the enantiomeric forms is that with plane polarised light. The ability of a solution to rotate plane polarised light is termed optical activity. The solutions of the two separate enantiomeric forms rotate plane polarised to the same degree but in opposite directions. So if a 1 M solution of one of the enantiomers rotated light by 30° in a clockwise direction, then a 1 M solution of the other enantiomer would rotate light by 30° in an anti clockwise direction.

This phenomenon was discovered by Louis Pasteur who laid the foundations in the understanding and separation of enantiomers. Pasteur was working with the

organic acids which crystallised during the fermentation of wine by yeast. The first acid to crystallise was Tartaric acid, which rotated plane polarised light in a clockwise direction. In the latter stages of crystallisation process a second acid, the so called racemic acid, was produced. Racemic acid was chemically identical to tartaric acid but was optically inactive. Pasteur observed that the sodium ammonium salt of racemic acid gave rise to two different crystal forms which were mirror images of each other. Pasteur was able to manually separate the different crystal forms of racemic acid and to show that the solutions produced from them rotated polarised light in opposite directions. One solution gave the same rotation as sodium ammonium tartrate and the other an equivalent rotation in the opposite direction. Henceforth tartaric acid was an incomplete description and the prefixes (+) for clockwise rotation or (−) for anti clockwise rotation were required. Pasteur surmised that the optical inactivity of racemic acid was due to equal amounts of (+) and (−) tartaric acid. In addition he made the important suggestion that the optical rotation was produced by the three dimensional distribution of the atoms within the molecule.

The nomenclature from this early pioneering work has survived with terms such as racemic mix being used to describe a 50:50 mixture of the two enantiomers.

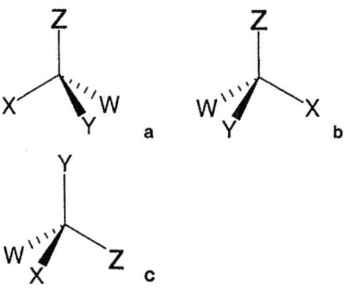

Figure 1.1. The two possible spatial arrangements of the groups W, X, Y, and Z around a tetrahedral carbon atom.

0009-5893/00/02   7-8 $ 03.00/0

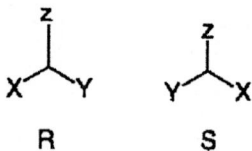

Figure 1.2. The R and S configurations.

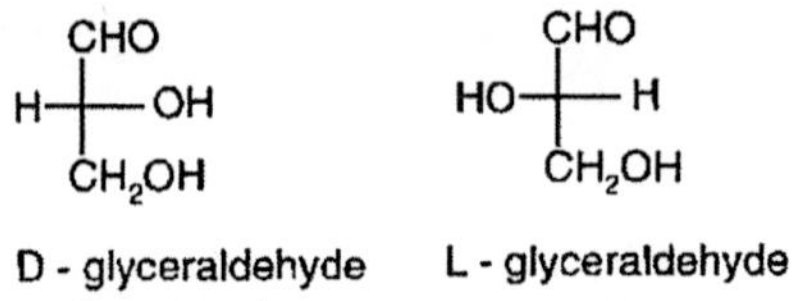

CHO
H——OH
CH₂OH

CHO
HO——H
CH₂OH

D - glyceraldehyde   L - glyceraldehyde

Figure 1.3. The D and L configurations of glyceraldehyde.

The alternative description system for the direction of rotation is that of dextrotatory or *d*, and laevorotatory or *l*. The routes of these terms are the Latin words *dexter* for right or right-hand and *laevus* for left or left-hand.

The terms *d* and *l*, and (+) and (–) only denote the direction in which plane polarised light is rotated and tell us nothing about the absolute stereochemical arrangement of the atoms in the molecule. The stereochemical arrangement about a particular asymmetric atom is normally described by using the Cahn-Ingold-Prelog rules [1]. The atoms which are attached to the asymmetric carbon atom are listed in order of decreasing priority according to their atomic mass. The asymmetric centre is viewed from opposite the atom of lowest priority. The remaining atoms will be distributed around the arc of a circle and are ordered in decreasing priority. If the remaining atoms are ordered in a clockwise direction the centre is designated (R), and if they are ordered in an anticlockwise direction the centre is designated (S). The terms (R) and (S) come from the Latin words *rectus* (right or correct) and *sinister* (left). If the atoms bound to the asymmetric centre are part of groups then the priority is established by considering the atoms further down the chain until a difference is apparent e.g. $CH_2OH > CH_3$.

As an example we can consider the pair of enantiomers depicted in Figure 1.1. Let us assume that the priority of the groups follows their position in the alphabet i.e. Z > Y > X > W. As W has the lowest priority we look at the asymmetric centre from the opposite side of the molecule from W. From this viewpoint the configuration shown in Figure 1.1a has the appearance shown in Figure 1.2a and the configuration shown in Figure 1.1b that shown in Figure 1.2b.

In Figure 1.2a the direction of decreasing precedence (Z–Y–X) is clockwise so the configuration is (R), and in Figure 1.2b the direction of precedence is anticlockwise so the configuration is (S).

For some natural products such as sugars the configuration at the asymmetric centre is often described by the use of the letters D and L instead of R and S. The D and L description system was proposed by Emil Fischer in 1919, years before X-ray techniques made it possible to determine the absolute configuration. The best that could be achieved in 1919 was a system that described the configuration relative to a given standard. Fischer chose (+) glyceraldehyde as the standard and guessed it (correctly) to have the configuration which he designated D. The (–) enantiomer of glyceraldehyde was given the designation L. All molecules, such as the larger sugars, which could be chemically related to (+) glyceraldehyde were also designated D. In order to use the system the molecule must be drawn in Fischer projection. In Fischer projection the sugar is drawn as a planar molecule with the CHO group at the top, the CH2OH group at the bottom and the H and OH groups at the sides. In this convention it is assumed that the central atom lies in the plane of the page and that groups which are drawn at the sides lie above the plane. The groups drawn above and below the central atom lie behind the plane of the page. With D glyceraldehyde the OH group lies to the right and the H to the left. The D and L configurations of glyceraldehyde are shown in Figure 1.3.

As was discussed above any compound which could be chemically related to (+) glyceraldehyde was given the designation D, regardless of the direction in which it rotates plane polarised light.

The D and L system suffers from confusion with the prefixes *d* and *l*, which are used to describe the direction of rotation, and has been largely replaced by the Cahn, Ingold, and Prelog system and the descriptors R and S. Usage of D and L is limited to areas such as amino acid and carbohydrate chemistry where trivial names are traditional and much easier to use.

The absolute configuration of the salts of tartaric acid was established by Bijvoet, Peerdeman, and Bommel who used a special single wavelength X-ray technique to distinguish between the two relative configurations [2]. The configuration of *d* sodium ammonium tartrate was established by replacing the ammonium ion with rubidium and employing zirconium K_α rays which just excite the rubidium atom. This process produces a phase lag and so the diffraction intensities are different for the two configuration.

1.2 Enantiomers and Living Organisms

1.2.1 Background

In recent years there has been a large increase in research on the synthesis and analysis of asymmetric compounds. The increase in research activity has arisen because of the growing realisation that the biological impact of one enantiomer can be significantly different to that of its partner.

The link between biochemical differences and different enantiomers is not surprising given that many living organisms favour one enantiomer over the other. Many of the basic building blocks of life, such as sugars and amino acids, have one or more asymmetric centres. It is common for the enantiomers which arise from the asymmetry to be produced and used to very different extents. With sugars the D-enantiomers predominate and with the amino acids the L-enantiomers predominate. The use of asymmetric building blocks means that the complex species assembled from them, such as enzymes and protein receptors can often be affected differently by different enantiomers.

Living organisms are highly efficient and effective at asymmetric synthesis and therefore both their products and their production machinery are valuable to the chemist in the synthesis and separation of enantiomers. Living organisms are the source of a large part of the so called "chiral pool" and enzymes and other proteins are increasingly used in synthesis.

A simple example of the different impacts of the enantiomers on living organisms can be found in their different smells. An example is that of (–) limonene which has a different smell from (+) limonene. If different enantiomers can have different smells it is not surprising that more profound physiological differences can also be caused. A case of more profound differences is that seen with the β-blocker propranolol (Figure 1.4). The racemic mixture was compared with the (+) enantiomer by determining its influence on the exercise tolerance of patients with angina pectoris. The racemate increased the length of time that the patients were able to exercise and reduced their heart rate whilst the (+) form showed no difference from that of the control [3].

The importance consequences of the use pharmaceutical agents with asymmetric centres was strongly emphasised by E.J. Ariens [4] who concluded that: 'Too often, and even without it being noticed, data in the scientific literature on mixtures of stereoisomers, racemates, are presented as if only one compound were involved. This neglect of stereochemical aspects of drug action, including metabolism, excretion etc. notwithstanding computerized curve fitting, generation of extensive tables with pharmacokinetic constants and postulation of complex multicompartment systems, degrades many pharmacokinetic studies to expensive "highly sophisticated scientific nonsense"'. Ariens gave examples where only one of the enantiomers produces the desired therapeutic effect [4]. The other enantiomer might be totally inert, have a different biological action, produce side effects, or even have the opposite effect to that of the first enantiomer.

The fact that the two enantiomers of a compound can produce different effects can lead to the temptation to simplify the situation by characterising the two enantiomers as 'good' and 'bad'. An example of this temptation can be seen in some of the discussion concerning the sedative Thalidomide (Figure 1.5) which was marketed as a racemate.

Thalidomide was marketed as a sedative in the early 1960s but in some cases was found to cause foetal abnormalities when it was prescribed for morning sickness. According to a commonly expressed view the toxicity arises from the (S)-$(-)$ form thalidomide alone. The evidence however does not seem to be clear cut and different workers obtained different results as summarised by De Camp [5]. More recently thalidomide enantiomers have been shown by Knoche and Blaschke to racemise quickly in various aqueous media [6]. The (S)-$(-)$ form had been found previously to be completely racemised after 2 hours in Rabbits. Further *in vitro* experiments indicated that plasma proteins catalyse the racemisation, and that in their presence the two enantiomers racemise at different rates. The half life of racemisation was measured in phosphate

buffer at pH 7.4; in Human and Rabbit citric plasma; and in different human serum albumin fractions. In phosphate buffer the half lives were the same within experimental error at 288.8 min and 260.5 min for the (+) and (–) forms respectively. In the plasma samples much higher racemisation rates were seen, with half lives of 11.5 and 8.3 min in human plasma and 9.3 and 6.5 in rabbit plasma. In defated human serum albumin the rates of thalidomide racemisation were similar to those seen in human plasma. In a fraction of albumin with a high content of fatty acids the racemisation rates were reduced. This and other results imply that a hydrophobic interaction between the thalidomide and the albumin is important in racemisation. Incubation of thalidomide with rat liver microsomes gave only low rates of racemisation. If racemisation occurs very rapidly *in vivo* it is very difficult to prove that the toxicity is only due to one of the enantiomers.

Metabolic inversion has also been shown to occur with a number of other different enantiomeric pharmaceutical compounds. The 2-arylpropionic acids are an important group of non-steroidal anti inflammatory drugs and their activity is believed to arise from the inhibition of prostaglandin synthesis. Whilst the *in vitro* activity of the (S) 2-arylpropionic acids is much greater than that of the (R) forms the differences seen *in vivo* are however much smaller. With ibuprofen (shown in Figure 1.6) for example the ratio of activity (S/R) is 160 *in vitro* but only 1.3 *in vivo*. The explanation for the differences is the metabolic inversion which occurs with the (R) enantiomers of at least some of the acids [7]. The *in vivo* conversion of (R) to (S) is thought to proceed via the formation of coenzyme A thioester intermediate [7, 8].

Metabolic inversion has also been shown to take place with mandelic acid [9]. Rats which were administered with the (S) mandelic acid excreted 80% as the (R) form and 20% as the (S) form. Rats which were dosed with (R) mandelic acid or the racemate also excreted the (R) enantiomer alone.

In general enantioselectivity for a drug or other compound by an organism can be divided into two classes: substrate enantioselectivity and product enantioselectivity [10]. Substrate enantioselectivity involves different metabolism of the enantiomers because of different binding affinities or different catalytic activity. Pro-

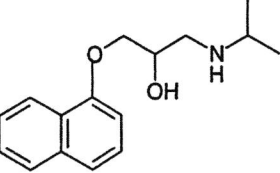

Figure 1.4. The structure of the β-blocker propranolol.

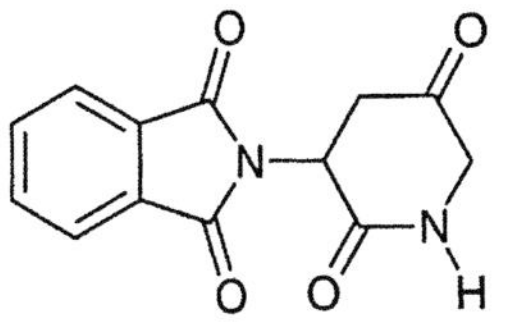

Figure 1.5. The structure of the sedative thalidomide.

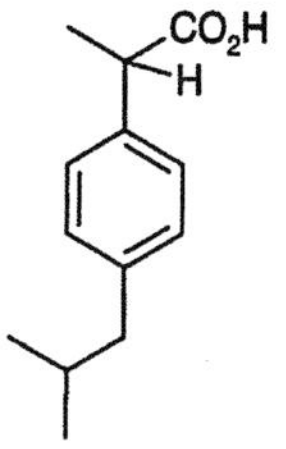

Figure 1.6. The structure of the anti-inflammatory ibuprofen.

duct enantioselectivity is the differential formation of enantiomeric metabolites from a common substrate. The difference in substrate enantioselectivities is often explained using the analogy of a hand fitting into a glove. The two enantiomers are compared with the right and left hands and the enzyme or other active site to the glove. Clearly one of the hands can (depending on the quality of the glove) fit much better than the other and so will have a higher binding affinity. Differences in catalytic activity can arise because a tighter fit can mean that inter-atomic distances and so activation energies are lower.

1.2.2 Regulatory Aspects

As detailed in the examples above, different enantiomers can produce both different pharmacokinetic (e.g. rates of metabolism) and pharmacodynamic (e.g. different receptor affinities) behaviour. Because of the potential for these differences analytical methods which are capable of resolving the enantiomers are vital for use in the toxicity and clinical testing of the drug. This requirement has been summarised by De Camp [5] "Good science requires that our conclusions are based on

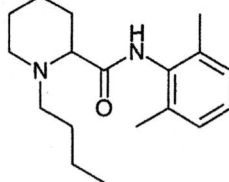

Figure 1.7. The structure of the anaesthetic bupivacaine.

experimental evidence that is derived from well-planned experiments. Such a level of planning should not neglect the potential for differences in properties for enantiomers of a chiral molecule in a chiral environment".

Suitable methods are also required by the regulatory authorities such as the US FDA (Food and Drug Administration) for the control of both the bulk drug and formulated product [11]. For an identity test to be specific it must be capable of discriminating between the enantiomers or between an enantiomer and the racemate. Purity and assay methods must also be capable of discriminating between enantiomers so that any batch to batch variations can be controlled for quality purposes.

Because of the recognition of the potential problems caused by compounds with an asymmetric centre many pharmaceutical companies have taken the decision to market all new drugs with such a centre as single enantiomers. The need for large quantities of enantiomerically pure materials for use as pharmaceuticals and as agrochemicals has given great emphasis to the development of large scale stereo selective synthesis [12, 13]. This emphasis has lead to the need to characterise a wide range of enantiomerically pure raw materials and intermediates in addition to the final drug product and marketed formulation.

1.2.3 Chiral Switches

The recognition that unwanted side effects can sometimes be reduced by the administration of a single enantiomer alone has lead to the re-evaluation of existing medicines. For drugs where the activity is associated with one enantiomer and significant toxicity with the other there is a case to replace the racemate with a single enantiomer. There are several examples of these so called chiral switches and some companies have specialised in their development.

An example is that of the local anaesthetic bupivacaine (Figure 1.7) which historically has been marketed as a racemate. Data indicate that the toxicity of the (R)-(+) enantiomer is significantly higher than that of the (S)-(–) enantiomer and that both enantiomers have similar efficacy. Work on isolated rabbit hearts showed that occurrence of severe arrhythmias was much less pronounced in hearts receiving the (S)-(–) enantiomer than those receiving the (R)-(+) enantiomer or the racemate [14]. Studies comparing the clinical efficacy of (S)-(–) bupivacaine and the racemate on patients undergoing hand surgery did not show any differences in performance [15].

1.2.4 Enantiomeric Agrochemicals

Differences in efficacy have also been observed for the different enantiomers of some herbicides, insecticides and other agrochemicals. An example is that of the synthetic pyrethoids. For example the (1R)-(+) trans enantiomer of resmethrin shows a much greater insecticidal activity than that of the (1S)-(–) trans enantiomer, although the toxicity to mice is similar [16]. Decisions on whether to market agrochemicals as single enantiomers or racemates are complex and depend on factors such as the activity and side effects of the enantiomers and production costs. There are several cases of different classes of agrochemicals being marketed as single enantiomers [13, 16]. For example the (R) enantiomer of the herbicide dichloroprop (2-(2,4-dichlorophenoxy) propionic acid) which shows higher activity.

1.3 The Separation of Enantiomers

Because of their identical behaviour in most physical and chemical systems, enantiomers are difficult to separate. Pasteur was able to physically separate the enantiomers of tartaric acid as they formed crystals which are mirror images of each other. This feature of different enantiomers forming different crystals is very rare and therefore physical separation is only possible in a limited number of cases. Most compounds will give crystals containing an equal number of the two enantiomers rather than different crystals each containing a single enantiomer.

One of the principle classical methods used for the separation of enantiomers is via the formation of diastereoisomers. Diastereoisomers is the description used to cover all other stereoisomeric forms of a compound other than the enantiomeric ones. Diastereoisomers have different physical and chemical properties and so are much easier to separate than enantiomers. Enantiomeric acids for example can be separated by the fractional crystallisation of the salts formed by reaction with an optically pure base (and vice versa). This traditional resolution method is widely used as crystallisation is a highly efficient large scale purification technique.

1.3.1 Chromatographic Separation of Enantiomers

In the past researchers only had limited tools to characterise the purity of the enantiomers they had prepared such as melting point and optical rotation measurements. Chromatographic and electrophoretic separation methods were developed because of the need to characterise chiral compounds to greater extents and with greater accuracy and precision. These separation methods have meant that it is possible to measure small amounts of the minor enantiomer in compounds of high optical purity. Separation methods have also helped to distinguish between chemical and enantiomeric purity. In addition chromatographic methods in particular have been applied to the separation of both enantiomers from a racemate in sufficient quantities that testing can be undertaken.

Chromatographic and electrophoretic resolutions of enantiomeric pairs are brought about via the permanent or transient formation of diastereoisomeric species. In the past the main emphasis has been on chromatographic separation methods and chiral HPLC, GC, TLC and SFC have all been employed. Most of the chromatographic procedures involve the use of bonded stationary phases containing a chiral selector. The alternative approach of adding a chiral selector to the mobile phase and using a non chiral stationary phase has also been used. With the chiral stationary phase approach one of the enantiomers forms a stronger interaction with the chiral selector in the stationary phase and so is more strongly retained. Chiral mobile phase additives function via the formation of temporary diastereoisomeric complexes in solution. Differences in either the stability or retention of the two complexes can lead to separation of the enantiomers.

A wide range of chiral selectors have been used, many of them being natural products. Some examples are: chiral ion pair reagents, chiral ligand exchange reagents, proteins such as serum albumins and α_1-acid glycoprotein, cyclodextrins and cellulose derivatives, alkaloids, modified amino acids, and macrocyclic antibiotics. In some cases the same chiral selector has been used in many different chromatographic techniques. The most widely used chromatographic technique for the separation of enantiomers is HPLC. Because of the great number of separation modes available (reversed phase, normal phase, ion pair etc.) HPLC can deal with compounds of very wide ranging polarity. HPLC is widely used in the analysis of enantiomeric pharmaceutical compounds. The use of GC and of SFC in the analysis of pharmaceutical agents is more limited due to the volatility (GC) and polarity (SFC) of the analyte.

In an interesting development Schurig and co-workers used a cyclodextrin coated capillary to perform 'unified enantioselective capillary chromatography' [17]. A 1 m × 50 μm fused silica capillary coated with permethylated-β-cyclodextrin was used to separate hexobarbital enantiomers by GC, SFC, and CE. The GC separation used hydrogen as the mobile phase, the SFC separation carbon dioxide, and the CE separation Tris-HCl buffer.

Whilst HPLC has been a tremendously successful technique for the separation of enantiomers there have been a number of problems which have lead researchers to examine alternative approaches. Some of the columns give only poor efficiencies, lack robustness and have long equilibration times. Many of the stationary phases and mobile phases are expensive leading to high running costs. In addition analysts with new enantiomers are faced with an almost bewildering array of HPLC stationary phases. Analysts are unsure how the structure and properties of their compound leads to a rational choice of stationary and mobile phase starting conditions.

1.4 Electrophoretic Separation of Enantiomers

1.4.1 The Development of Capillary Electrophoresis

The wide spread use of electrophoresis to perform separations of enantiomers is more recent than the use of chromatography. This is partly because the development of a capillary or tubular separation format (with attendant advantages in automation, detection and speed) is more recent in electrophoresis. Traditionally electrophoresis has been carried out in a planar format and problems with thermal gradients have required the use of low electric field strengths and support media such as gels or paper. Whilst gel electrophoresis has been widely and very successfully used in the analysis of biomolecules such as DNA and proteins, the application to small molecules has been limited.

In the absence of supporting media such as gels, convection currents and thermal gradients lead to broadening of the analyte bands and hence a reduction in resolution. One way around these problems is to carry out the separation in narrow tubes [18]. With tubes having a narrow cross sectional the heat generated by the passage of an electrical current across a large potential difference can be effectively removed [19].

In 1979 Mikkers, Everaerts and Verheggen published work on zone electrophoresis in 200 μm PTFE tubes [20]. They used buffer systems such as acetic acid/histidine to separate a range of organic acids and inorganic anions and employed both conductivity and UV detection. They demonstrated the benefit of using a sample solvent of a lower ionic strength than the buffer and that the detector response was directly proportional to sample concentration. In a separate theoretical paper the same workers showed that asymmetric analyte peaks were caused by electrophoretic mobility differences between the analyte and the buffer ions [21]. The analyte peak can either front or tail according to whether the analyte has a higher or lower electrophoretic mobility than the buffer ion. This dispersion effect can be reduced by a better matching of mobilities or by the use of a sample concentration which is much lower than that of the buffer ions.

The current practice of capillary electrophoresis can be traced to the pioneering work of Jorgenson and Lukacs in the

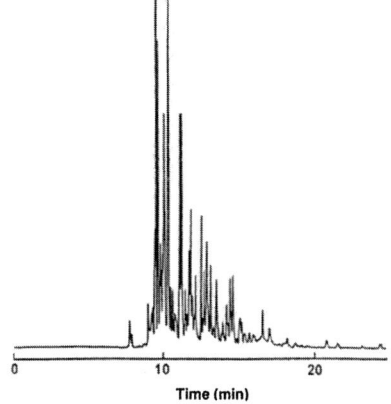

Figure 1.8. Electropherogram of fluorescamine labeled peptides obtained from a tryptic digest of chicken albumin. Reprinted from [23] with permission from Elsevier Science.

early 1980s [22–26]. A number of important technical innovations were introduced and several separations which indicated the potential of the technique were shown. Electrophoresis was carried out in glass capillaries with an internal diameter of 75 μm and lengths of up to 1 metre: Voltages of up to 30 kV were employed along with sensitive on-column fluorescence detection. Fluorescence was used because of problems with the sensitivity of UV absorbance detection.

Jorgenson and Lukacs showed separations of dansylated amino acids and used fluorescamine to derivatise alkylamines and peptides. The potential of the technique in biochemistry was shown by the analysis of proteins and the peptides produced by tryptic digests. The separation of the fluorescamine derivatives of the amines in a urine sample can be seen in Figure 1.8. The sharpness of the peaks and large number of sample components shows the performance of the technique.

The relationship between the efficiency and the applied voltage was investigated and plate counts of up to 400,000 were obtained. It was also shown that resolution could be improved by a reduction in the electroosmotic mobility. In another experiment Capillary Electro Chromatography (CEC) was performed by packing a 170 μm × 58 cm capillary with 10 μm silica particles coated with a C18 stationary phase. Perylene and 9-methylanthracene were separated with plate counts of 23,000 and 31,000 respectively. In their conclusions [23] the authors noted: 'The greatest obstacle to further development and utilization of capillaries is the requirement of extremely sensitive detectors, and more

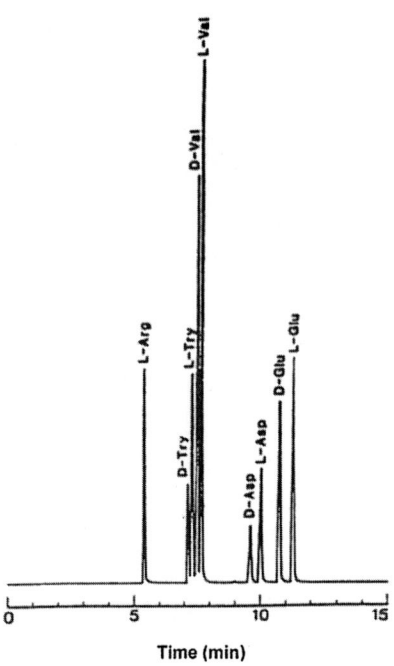

Figure 1.9. Electropherogram of a mixture of four DNS-DL-amino acids. Reprinted from [28] with permission of American Chemical Society.

types of detectors with higher sensitivity are greatly needed. A better understanding of capillary surface modification will also be important, both for improved capillary surface deactivation and for better control over electroosmotic flow'. These observations are arguably as true now as they were nearly twenty years ago.

1.5 Enantiomer Separations by Capillary Electrophoresis

1.5.1 Ligand Exchange

The chiral selectors investigated in CE are those which had already been employed in HPLC. Some of the earliest enantiomer separations by CE were achieved via ligand exchange with other enantiomeric species complexed to copper (II) ions [27, 28]. In 1985 Gassman, Kuo, and Zare [27] were able to separate racemic mixtures of several dansylated amino acids by the use of a L-histidine copper (II) complex. The derivatised amino acids enantiomers are thought to exchange with one of the L-histidine ligands to form diastereoisomeric species of different stabilities. Amino acids bound to the copper complex migrate faster than the unbound ones because of the positive charge on the copper complex. Replacing L-histidine with D-histidine lead to a reversal in the migration order.

The use of racemic histidine resulted in a loss in enantiomeric resolution (although the different amino acids were still resolved from each other). The separations were carried out in a 75 μm × 75 cm fused silica capillary and the use of dansyl derivatives meant that a He-Cd laser and fluorescence detection could be employed.

In a subsequent paper better separations of the dansylated amino acid enantiomers were obtained by replacing histidine with aspartame [28]. Eighteen amino acids were examined and the enantiomers of fourteen could be resolved with the D enantiomer forming a more stable complex in each case. The electropherogram of some of the amino acids is shown in Figure 1.9.

The use of micelles formed by an anionic surfactant, sodium tetradecyl sulphate, enabled the separation between some of the neutral amino acids to be improved. Temperature control was achieved by surrounding part of the capillary with a teflon tube and pumping a liquid coolant through it. The buffer pH, temperature, and aspartame:copper ratio were all varied to determine the influence on resolution. With some amino acids the resolution was little influenced by changes in these parameters but with others large changes were seen. With dansyl DL valine a plate count of 130,000 was obtained and this value did not change as the temperature was varied between 1 °C and 40 °C.

1.5.2 Cyclodextrins

The use of cyclodextrins as chiral selectors in CE grew out of their use as a way of separating achiral isomeric species. In 1985 for example Terabe and co-workers used ionic β-cyclodextrin derivatives to separate the isomers of cresol and xylidine and other aromatic compounds [29].

Guttman and co-workers employed α, β, and γ-cyclodextrins which had been incorporated into a cross linked polyacrylamide gel [30]. The gel filled capillaries were used to separate the enantiomers of several dansylated amino acids. The β-cyclodextrin was found to give the best results and this was ascribed to a better match between the sizes of the amino acids and the cyclodextrin cavity. A number of parameters such as the cyclodextrin concentration, temperature, and electrical field strength were varied and their affect on the chiral selectivity determined. The authors also determined the affect of add-

ing methanol to the buffer system and showed it to have a strong influence on resolution. The enantiomers of twelve dansylated amino acids were resolved and separation efficiencies varied between 50,000 and 100,000 theoretical plates. In addition equations were developed to model enantiomeric selectivity as a function of the cyclodextrin concentration.

The use of cyclodextrins dissolved in the separation buffer was also investigated by several workers. Fanali used β-cyclodextrin and 2,6-di-o-methyl-β-cyclodextrin to separate the enantiomers of ephedrine, norephedrine, epinephrine, norepinephrine and isoproterenol [31]. Whilst 20 mM β-cyclodextrin gave only poor resolution of the enantiomers of ephedrine and isoproterenol, baseline resolution of all of the enantiomeric pairs was achieved with 18 mM 2,6-di-o-methyl-β-cyclodextrin. In each case the (+) isomer has the higher affinity for the cyclodextrin. The separation of the enantiomers was shown to increase with increasing cyclodextrin concentration. Experiments with different concentrations of tri-o-methyl-β-cyclodextrin did not give any resolution of the enantiomers.

In other work the enantiomers of tryptophan were separated using α-cyclodextrin [32]. The resolution was shown to be a function of the cyclodextrin concentration and the type of acid used to prepare the buffer. The separation obtained by using 40 mM α-cyclodextrin in a 100 mM phosphate buffer at pH 2.5 is shown in Figure 1.10.

A buffer containing 20 mM 2,6-di-o-methyl-β-cyclodextrin was used to determine the enantiomeric purity of commercial samples of (−) epinephrine. Buffers containing α, β, and γ-cyclodextrin and di-o-methyl, and tri-o-methyl-β-cyclodextrin were used to try and separate the enantiomers of terbutaline and propranolol [33]. The degree of separation was shown to be strongly affected by both the type and concentration of the cyclodextrin employed. The separation was also influenced by the addition of methanol to the buffer. The enantiomers of chloramphenicol and thioridazine were separated by the use of 2,6-di-o-methyl-β-cyclodextrin and γ-cyclodextrin respectively [34]. The separation of the chloramphenicol enantiomers was improved by the addition of a substituted cellulose to the buffer.

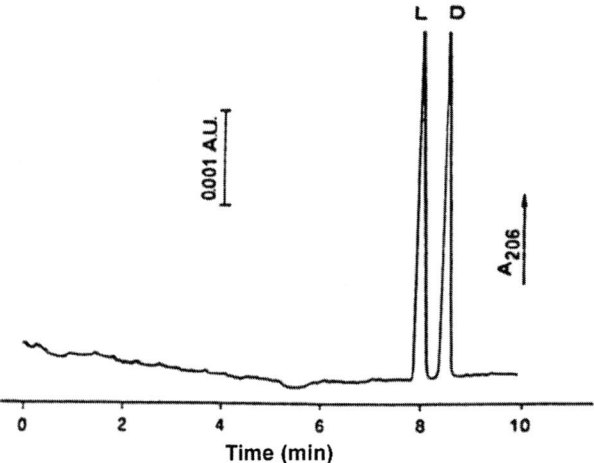

Figure 1.10. Analysis of a racemic mixture of DL-tryptophan. Reprinted from [32] by permission of WILEY-VCH.

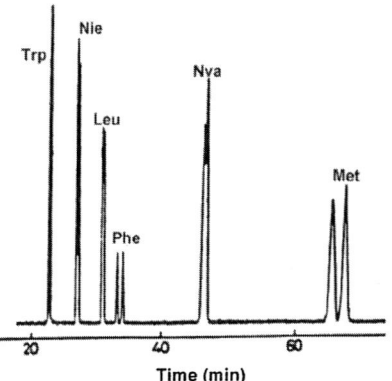

Figure 1.11. Micellar EKC separation of DNS-DL-amino acids. Reprinted from [35] with permission from Elsevier science.

1.5.3 Chiral Surfactants

The use of surfactants to separate enantiomers developed from the concept of using ionic surfactants to separate neutral molecules in CE. In Micellar Electro Kinetic Chromatography (MEKC) neutral analytes are separated from each other on the basis of their differing affinities for micelles formed from ionic surfactants such as Sodium Dodecyl Sulphate (SDS). With SDS micelles the principle interaction controlling the affinity is hydrophobicity. In order for the micelles to separate enantiomers the micelles have to have an asymmetric element such that the affinities are different.

Terabe, Shibata, and Miyashita used micelles formed from bile salts to separate the enantiomers of the dansylated amino acids tryptophan, norleucine, leucine, norvaline, methionine, and phenylalanine [35]. The degree of separation varied with the amino acid and the bile acid used. The enantiomers of phenylalanine and methionine were baseline resolved whereas a partial separation only was seen for the others. The best separations were obtained with sodium taurodeoxycholate. Figure 1.11 shows the separation of the dansylated amino acids using a buffer containing 50 mM taurodeoxycholate and 50 mM phosphate at pH 3.0.

Bile salt micelles were also used to separate the enantiomers of several binaphthyl compounds [36]. Sodium deoxycholate was used to separate the enantiomers of 1,1'-bi-2-naphthol and 1,1'-binaphthyl-diyl-hydrogen-phosphate. Sodium taurodeoxycholate was used to separate the en-

antiomers of 1,1'-binaphthyl-dicarboxylic-acid.

Micelles formed from SDS and sodium N-dodecanoyl-L-valinate were used to separate the benzoyl, 4-nitrobenzoyl, and 3,5-dinitrobenzoyl derivatives derivatives of the enantiomers of alanine, valine, leucine, and phenylalanine [37]. The 3,5-dinitrobenzoyl derivatives were found to give the best resolution and the L enantiomers were found to have higher affinities for the micelle than the D enantiomers. The individual amino acid derivatives were separated from each other on the basis of differences in their hydrophobicities alanine had the lowest affinity for the micelle and phenylalanine the highest.

1.5.4 Mixed Cyclodextrin and Surfactant Systems

Buffer systems containing both a cyclodextrin and a surfactant were developed to extend separation to neutral enantiomers. An example is the use of cyclodextrins and SDS by Nishi, Fukayama, and Terabe to separate the enantiomers of some barbiturates and other compounds [38]. The authors examined the selectivity of α, β, and γ-cyclodextrin and di-o-methyl, and tri-o-methyl-β-cyclodextrin. The γ-cyclodextrin showed enantiomeric selectivity for the widest range of compounds and α-cyclodextrin no selectivity at all. The selectivity was shown to vary with the concentration of γ-cyclodextrin and with the addition of 10% methanol to the buffer. It was also found that the selectivity could be altered by the addition of sodium d-cam-

phor-10-sulphonate or sodium l-menthoxyacetate to the buffer.

Buffer systems containing SDS and either β-cyclodextrin or γ-cyclodextrin were used to separate the enantiomers of some amino acids which had been derivatised using naphthalene-2,3-dicarboxaldehyde in the presence of cyanide [39]. Figure 1.12 shows the separation of the enantiomers of five of the amino acids using a buffer containing 10 mM γ-cyclodextrin, 50 mM SDS, and 100 mM borate at pH 9.0.

1.6 Conclusion

The study of enantiomers is important because of the potential for differences in the changes they can produce in living systems. Analytical methods which separate enantiomers are tools which can make a major contribution to our understanding of the differences in their behaviour. Because of the potential for differences in biological affects, enantiomer separation methods are an important part of the tests required to ensure the consistency of pharmaceutical compounds. CE is an important additional tool for scientists interested in studying enantiomers and ensuring their quality and safety.

References

[1] Cahn, R.S.; Ingold, C.; Prelog, V. Specification of Molecular Chirality, *Angew. Chem. Internat. Edit.* **1966**, *5*, 385–415.

[2] Bijvoet, J.M.; Peerdeman, A.F.; van Bommel, A.J. Determination of the absolute

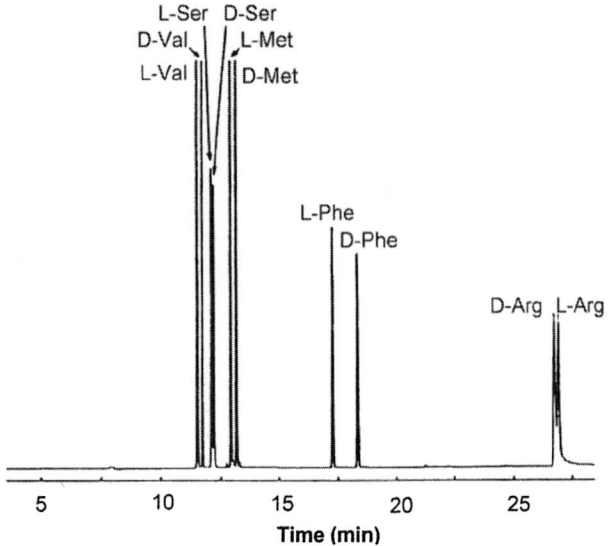

Figure 1.12. Electropherogram of a mixture of five CBI-DL-amino acids. Reprinted from [39] with permission of American Chemical Society.

configuration of optically active compounds by means of X-rays, *Nature* **1951**, *168*, 271–272.

[3] Wilson, A.G.; Brooke, O.G.; Lloyd, H.J.; Robinson, B.F. Mechanism of action of β-Adrenergic Receptor Blocking Agents in Angina Pectoris: Comparison of Action of Propranolol with Dexpropranolol and Practolol, *Br. Med. J.* **1969**, *4*, 399–401.

[4] Ariëns, E.J. Stereochemistry, a Basis for Sophisticated Nonsense in Pharmacokinetics and Clinical Pharmacology, *Eur. J. Clin. Pharmacol.* **1984**, *26*, 663–668.

[5] De Camp, W.H. The FDA Perspective on the Development of Stereoisomers, *Chirality* **1989**, *1*, 2–6.

[6] Knoche, B.; Blaschke, G. Investigations on the *in vitro* racemization of thalidomide by high-performance liquid chromatography, *J. Chromatogr. A* **1994**, *666*, 235–240.

[7] Caldwell, J.; Hutt, A.J.; Fournel-Gigleux, S. The metabolic chiral inversion and dispositional enantioselectivity of the 2-arylpropionic acids and their biological consequences, *Biochem. Pharmacol.* **1988**, *37*, 105–114.

[8] Tracy, T.S.; Hall, S.D. Metabolic inversion of (*R*) Ibuprofen, Epimerization and hydrolysis of ibuprofenyl-coenzyme A, *Drug Metab. Dispos.* **1992**, *20*, 322–327.

[9] Drummond, L.; Caldwell, J.; Wilson, H.K. The stereoselectivity of 1,2-phenylethanediol and mandelic acid metabolism and disposition in the rat, *Xenobiotica* **1990**, *20*, 159–168.

[10] Testa, B. Mechanisms of Chiral Recognition in Xenobiotic Metabolism and Drug-Receptor Interactions, *Chirality* **1989**, *1*, 7–9.

[11] De Camp, W.H. Chiral drugs: the FDA perspective on manufacturing and control, *J. Pharm. Biomed. Anal.* **1993**, *11*, 1167–1172.

[12] Collins, A.N.; Sheldrake, G.N.; Crosby, J. (eds.), in Chirality in Industry II. Developments in the Manufacture and Applications of Optically Active Compounds, J. Wiley & Sons, New York, **1997**.

[13] Sheldon, R.A. Chirotechnology: Industrial Synthesis of Optically Active Compounds, Marcel Dekker, New York, **1993**.

[14] Mazoit, J.X.; Boïco, O.; Samii, K. Myocardial Uptake of Bupivacaine: II. Pharmacokinetics and Pharmacodynamics of Bupivacaine Enantiomers in the Isolated Perfused Rabbit Heart. *Anesth. Analg.* **1993**, *77*, 477–482.

[15] Cox, C.R.; Checketts, M.R.; Mackenzie, N.; Scott, N.B.; Bannister, J. Comparison of *S*(–)-bupivacaine with racemic (*RS*)-bupivacaine in supraclavicular brachial plexus block. *Br. J. Anaesth.* **1998**, *80*, 594–598.

[16] Kurihara, N.; Miyamoto, J.; Paulson, G.D.; Zeeh, B.; Skidmore, M.W.; Hollingworth, R.M.; Kuiper, H.A. IUPAC Reports on Pesticides (37), Chirality in synthetic agrochemicals: bioactivity and safety consideration. *Pure & Appl. Chem.* **1997**, *69*, No. 9, 2007–2025.

[17] Jung, M.; Mayer, S.; Schurig, V. Enantiomer Separation by GC, SFC, and CE on immobilized Polysiloxane-Bonded Cyclodextrins, *LC-GC Int* **1994**, *7*, 340–347.

[18] Hjertén, S. Free zone electrophoresis, *Chromatogr. Rev.* **1967**, *9*, 122–219.

[19] Verheggen, Th. P.E.M.; Mikkers, F.E.P.; Everaerts, F.M. Isotachophoresis in narrow-bore tubes. Influence of the diameter of the separation compartment, *J. Chromatogr.* **1977**, *132*, 205–215.

[20] Mikkers, F.E.P.; Everaerts, F.M.; Verheggen, Th. P.E.M. High performance zone electrophoresis, *J. Chromatogr.* **1979**, *169*, 11–20.

[21] Mikkers, F.E.P.; Everaerts, F.M.; Verheggen, Th. P.E.M. Concentration distributions in free zone electrophoresis, *J. Chromatogr.* **1979**, *169*, 1–10.

[22] Jorgenson, J.W.; Lukacs, K.D. Zone Electrophoresis in Open Tubular Glass Capillaries, *Anal. Chem.* **1981**, *53*, 1298–1302.

[23] Jorgenson, J.W.; Lukacs, K.D. High-resolution separations, based upon electrophoresis and electroosmosis, *J. Chromatogr.* **1981**, *218*, 209–216.

[24] Jorgenson, J.W.; Lukacs, K.D. Free-Zone Electrophoresis in Glass Capillaries, *Clin. Chem.* **1981**, *27*, 1551–1553.

[25] Jorgenson, J.W.; Lukacs, K.D. Zone Electrophoresis in Open-Tubular Glass Capillaries: Preliminary Data on Performance, *J.H.R.C. & C.C.* **1981**, *4*, 230–231.

[26] Jorgenson, J.W.; Lukacs, K.D. Capillary Zone Electrophoresis, *Science* **1983**, *222*, 266–272.

[27] Gassman, E.; Kuo, J.E.; Zare, R.N. Electrokinetic Separation of Chiral Compounds, *Science* **1985**, *230*, 813–814.

[28] Gozel, P.; Gassman, E.; Michelsen, H.; Zare, R.N. Electrokinetic Resolution of Amino Acid Enantiomers with Copper (II)-Aspartame Support Electrolyte, *Anal. Chem.* **1987**, *59*, 44–49.

[29] Terabe, S.; Ozaki, H.; Otsuka, K.; Ando, T. Electrokinetic chromatography with 2-O-carboxymethyl-β-cyclodextrin as a moving 'stationary' phase, *J. Chromatogr.* **1985**, *332*, 211–217.

[30] Guttman, A.; Paulus, A.; Cohen, A.S.; Grinberg, N.; Karger, B.L. Use of complexing agents for selective separation in high-performance capillary electrophoresis Chiral resolution via cyclodextrins incorporated within polyacrylamide gel columns, *J. Chromatogr.* **1988**, *448*, 41–53.

[31] Fanali, S. Separation of optical isomers by capillary zone electrophoresis based on host-guest complexation with cyclodextrins, *J. Chromatogr.* **1989**, *474*, 441–446.

[32] Fanali, S.; Bocek, P. Enantiomeric resolution by using capillary zone electrophoresis: Resolution of racemic tryptophan and determination of the enantiomer composition of pharmaceutical epinephrine, *Electrophoresis* **1990**, *11*, 757–760.

[33] Fanali, S. Use of cyclodextrins in capillary zone electrophoresis Resolution of terbutaline and propranolol enantiomers, *J. Chromatogr.* **1991**, *545*, 437–444.

[34] Snopek, J.; Soini, H.; Novotny, M.; Smolkova-Keulemansova, E.; Jelinek, I. Selected applications of cyclodextrin selectors in capillary electrophoresis, *J. Chromatogr.* **1991**, *559*, 215–222.

[35] Terabe, S.; Shibata, M.; Miyashita, Y. Chiral separation by electrokinetic chromatography with bile salt micelles, *J. Chromatogr.* **1989**, *480*, 403–411.

[36] Cole, R.O.; Sepaniak, M.J.; Hinze, W.L. Optimization of binaphthyl enantiomer separations by capillary zone electrophoresis using mobile phases containing bile salts and organic solvent, *J.H.R.C. & C.C.* **1990**, *13*, 579–582.

[37] Dobashi, A.; Ono, T.; Hara, S.; Yamaguchi, J. Optical resolution of enantiomers with chiral mixed micelles by electrokinetic chromatography, *Anal. Chem.* **1989**, *61*, 1984–1986.

[38] Nishi, H.; Fukuyama, T.; Terabe, S. Chiral separation by cyclodextrin-modified micellar electrokinetic chromatography, *J. Chromatogr.* **1991**, *553*, 503–516.

[39] Ueda, T.; Kitamura, F.; Mitchell, R.; Metcalf, T.; Kuwana, T.; Nakamoto, A. Chiral separation of naphthalene-2,3-dicarboxaldehyde-labeled amino acid enantiomers by cyclodextrin-modified micellar electrokinetic chromatography with laser-induced fluorescence detection, *Anal. Chem.* **1991**, *63*, 2979–2981.

The Principles of Separation in CE

2.1 Introduction: Movement and Separation

In his book *Unified separation science* the late J. Calvin. Giddings, one of the great analytical chemists of the 20th century, defined separation as follows: "separation is the art and science of maximizing separative transport relative to dispersive transport" [1].

The aim of the analyst is thus to choose processes and conditions which allow the component bands to be moved apart whilst minimising the extent to which they become diffuse. Separation is not a trivial exercise as entropy, one of the fundamental driving forces in the universe, tends to make things mixed and dilute rather than separated and concentrated.

The choice by Giddings of the words art and science is interesting and perhaps reflects the fact that understanding of separation science is still incomplete. Success in the design and development of separative methods depends upon experience and imagination in addition to an understanding of the fundamentals.

Because of the need to consider the effects of both separative and dispersive transport it is normal for the two to be considered together in the concept of resolution. Resolution between two components depends not only on the average distance between the peaks or bands but also upon the spread of the individual peaks.

$$R_S = f(\text{Selectivity, efficiency}) \quad (2.1)$$

In simple terms the efficiency corresponds to the sharpness of the peaks or bands and the selectivity to the distance between them. These two features are illustrated by the simulation shown in Figure 2.1 which was obtained by modelling the two component peaks using a normal probability distribution.

Figure 2.1a corresponds to a system with the worst of both worlds: low selectivity and low efficiency. The low effi-

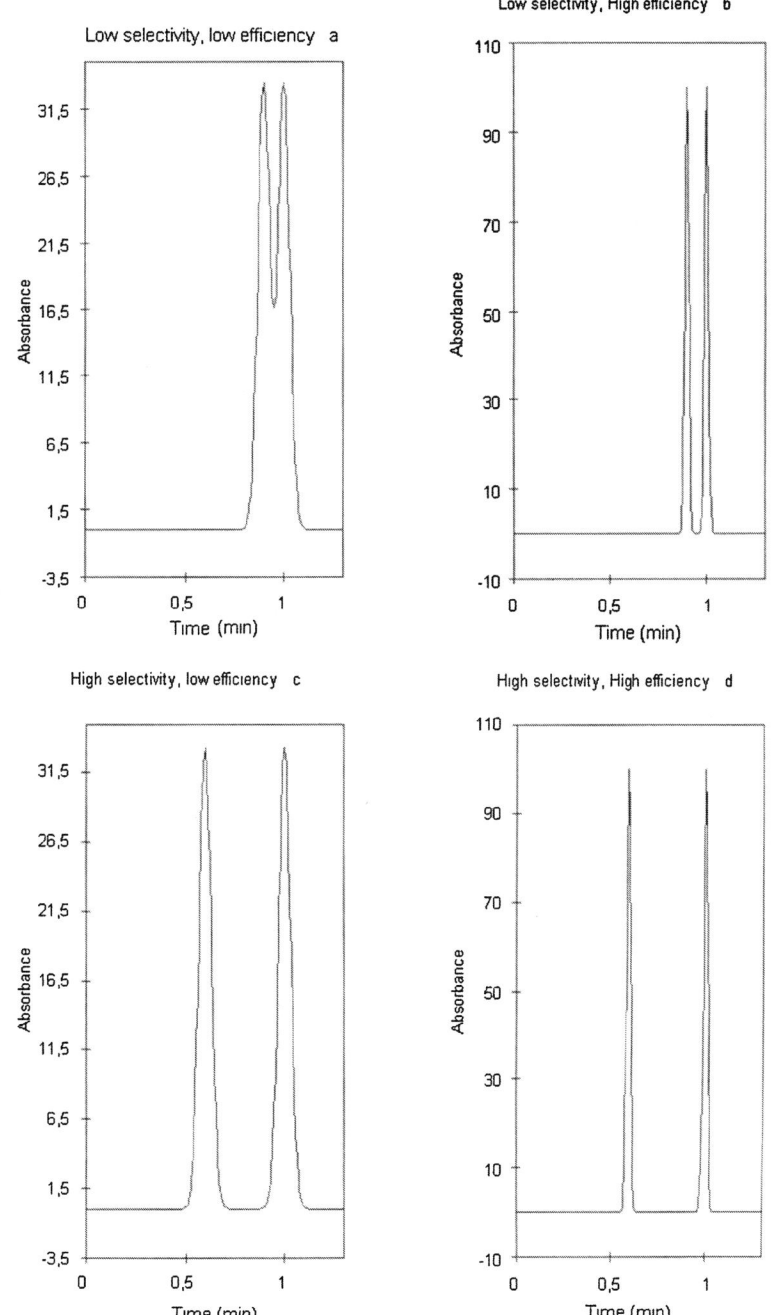

Figure 2.1. Simulated separations of enantiomers by CE showing the effects of separation selectivity and efficiency.

0009-5893/00/02   15-9 $ 03.00/0

ciency and selectivity results in poor resolution with considerable overlap of the two peaks. Low resolution is not a satisfactory situation as it leads to considerable uncertainty in the quantification of the two components. The errors will be greatest for samples where one component is present at a much lower level than the other, for example with an enantiomericaly pure pharmaceutical. The resolution between the two components can be improved by either increasing the efficiency or selectivity of the system. In Figure 2.1c the efficiency is the same as in 2.1a but the selectivity is four times higher. In Figure 2.1b the selectivity is the same as 2.1a but the efficiency is nine times higher. High separation efficiencies mean that systems with relatively low selectivity can still give rise to high resolution. Increasing the efficiency also has the benefit of increasing the height of the peak and so improving the detection limits. The ideal situation is represented by 2.1d where both the efficiency and selectivity are high, leading to high resolution.

The selectivity is a reflection of the ability of the system to discriminate between one enantiomer and another. This selectivity is a function of the chemistries of the system and the analytes. Because both system and analyte are involved the same selectivity cannot be expected for all enantiomers. In HPLC the selectivity arises from the functional groups in the stationary phase and the properties of the mobile phase. As the selectivity is different for different analytes no single stationary phase has proved entirely satisfactory for all classes of compounds. This situation has lead to the proliferation of a wide array of stationary phase chemistries relying on different types of interactions. This proliferation is in contrast to reversed-phase achiral HPLC where C18 stationary phase chemistries are widely used.

As CE is capable of producing much higher separating efficiencies than are common in HPLC it has the potential to give good resolution even when the selectivity is low. The promise of CE is therefore of systems with more universal applicability than may be possible with HPLC. More universal applicability has advantages to the analyst in terms of both simplifying method development and logistics.

2.2 Separative and Non-Separative Transport in CE

In CE the analytes move under the influence of an electrical field by two independent processes: electrophoresis and electroosmosis. The observed net movement of the analyte is the vector sum of these two transport processes.

Electrophoresis is the movement of charged species in an electrical field and depends upon the field strength and the size, charge, and shape of the analyte. Electrophoretic separation between two or more analytes therefore depends upon either inherent or induced differences in the properties of the analytes. Enhancing and exploiting differences in the size, charge or shape of the analytes is a key element in separating them by electrophoresis.

Electroosmosis is the bulk flow of buffer solution through the capillary and depends upon charge differences at the interface between the capillary wall and the buffer solution. As electroosmosis depends only upon the properties of the capillary surface and the buffer it is the same for all analytes. Because the electroosmosis experienced is the same for all analytes it is a non separative transport process. Electroosmosis is a very important as whilst it is non separative transport it alters the extent to which separative transport, electrophoretic mobility differences, can produce a separation.

The observed, or measured, velocity of an analyte is the vector sum of the electrophoretic and electroosmotic velocities and is given in equation (2.2).

$$U_{obs} = U_{eph} + U_{eo} \qquad (2.2)$$

Where U_{obs} is the observed velocity and U_{eph} and U_{eo} are the electrophoretic and electroosmotic velocities respectively.

As the electrophoretic and electroosmotic velocities are proportional to the strength of the applied electrical field it is normal to replace them with the corresponding proportionality constants μ_{eph} and μ_{eo}.

$$U_{obs} = E(\mu_{eph} + \mu_{eo}) \qquad (2.3)$$

Equation (2.3) can be re written as equation (2.4).

$$\frac{l}{t} = \frac{V}{L}(\mu_{eph} + \mu_{eo}) \qquad (2.4)$$

Where l is the length of the capillary from inlet to detector, L is the total length of the capillary, t is the analyte migration time, and V is the applied voltage.

The electroosmotic mobility for a particular capillary and buffer system can be determined by measuring the migration time of a neutral analyte.

The resolution between two enantiomers depends upon a number of parameters including those discussed above. Various equations have been proposed, amongst them equation (2.5) from Terabe [2] which is based upon earlier considerations of Giddings [3] and Jorgenson and Lukacs [4].

$$R_S = \left(\frac{V}{32D}\right)^{0.5} \cdot \left(\frac{l}{L}\right)^{0.5} \cdot \frac{\Delta\mu_{ep}}{(\mu_{ep} + \mu_{eo})^{0.5}} \qquad (2.5)$$

Where D is the diffusion coefficient of the analyte, $\Delta\mu_{ep}$ is the electrophoretic mobility difference between the two enantiomers, μ_{ep} is the average electrophoretic mobility of the two enantiomers, and μ_{eo} is the electroosmotic mobility. Equation (2.5) is based upon the simplifying assumption that band broadening (dispersive transport) is only caused by diffusion along the length of the capillary. All other potential causes of dispersion such as the injection process or thermal gradients are ignored.

2.3 Electrophoretic Mobility

The electrophoretic mobility of an analyte depends upon its size, charge, and shape. The importance of these factors can be considered by reference to examples of the behaviour of achiral analytes.

2.3.1 Molecular Size

Figure 2.2 shows the separation of pyridine from some 2-n-alkylpyridines (methyl, ethyl, propyl, pentyl and hexyl) using a 40 mM lithium phosphate buffer at pH 2.5. At pH 2.5 all of the pyridines are fully protonated and so the separation is based on size alone with pyridine having the highest electrophoretic mobility and 2-n-hexyl pyridine the lowest. The electrophoretic mobilities of the pyridines were correlated with their size using Offord's parameter [5], $q/M^{0.66}$ where q is the charge and M the molecular weight. Linear regression analysis gave an r^2 value of 0.999 [6].

Offord's parameter was used to produce similar correlations between electrophoretic mobility and size for oligoalanines and oligoglycines ranging in size from 2 to 6 amino units [7]. Other descriptions of molecular size have also been used to demonstrate successful correlations with electrophoretic mobility. For large molecules which have many charged centres such as polypeptides or proteins models of electrophoretic mobility are complex and many different models have been employed [8].

2.3.2 Molecular Charge

The molecular charge is related to the pK_a value of the acidic or basic function on the analyte and the pH of the buffer system. The simplest cases are those where the analyte is either monoacidic or monobasic. When the charge is limited to the range 0 to 1 it may be estimated via the degree of dissociation, α, of the molecule as shown in equation (2.6).

$$\alpha = \frac{1}{10^{(pH-pK_a)} + 1} \qquad (2.6)$$

For bases the size of the positive charge is given by α and for acidic molecules the size of the negative charge by $1-\alpha$. This procedure for estimating charge was used to determine the relationship between the measured electrophoretic mobility and the charge of the peptide diglycine [7]. The electrophoretic mobility was measured in six buffer systems covering the range pH 2.51 to 4.15. In this pH range the calculated net molecular charge varied between about 0.8 and 0.1. Plotting the measured electrophoretic mobility against the calculated charge resulted in a straight line shown in Figure 2.3.

Figure 2.3 shows a straight line with a correlation coefficient of 0.99 and a zero intercept. Similar linear relationships between calculated charge and measured electrophoretic mobilities were seen with nine other peptide oligomers [7] and with 2 and 4 methyl pyridine [9]. For analytes which have many acidic or basic groups such as polypeptides it may be possible to estimate the overall charge by summing the contributions from the individual groups. As the electrophoretic mobility is proportional to the calculated charge the manipulation of pH can be a very powerful tool in the separation of isomers. Equation (2.6) can be used to describe the charges on two isomers or species with dif-

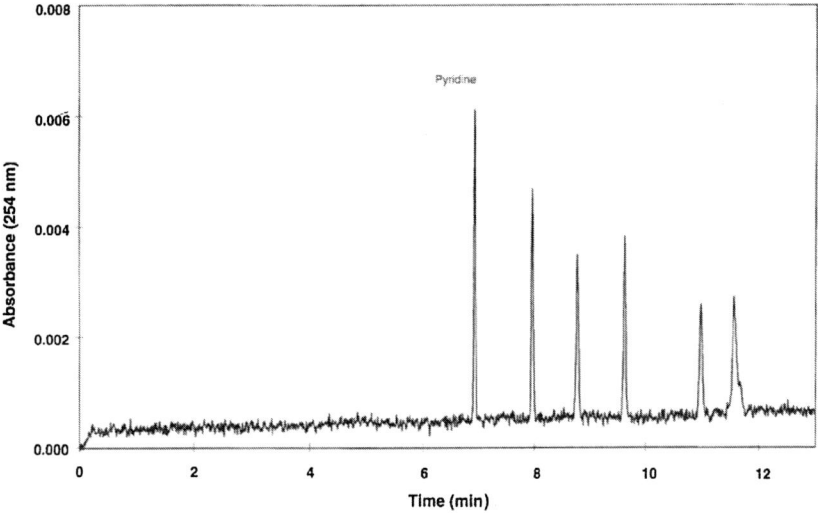

Figure 2.2. Electropherogram showing the separation of pyridine and some 2-*n*-alkyl-pyridines at pH 2.5.

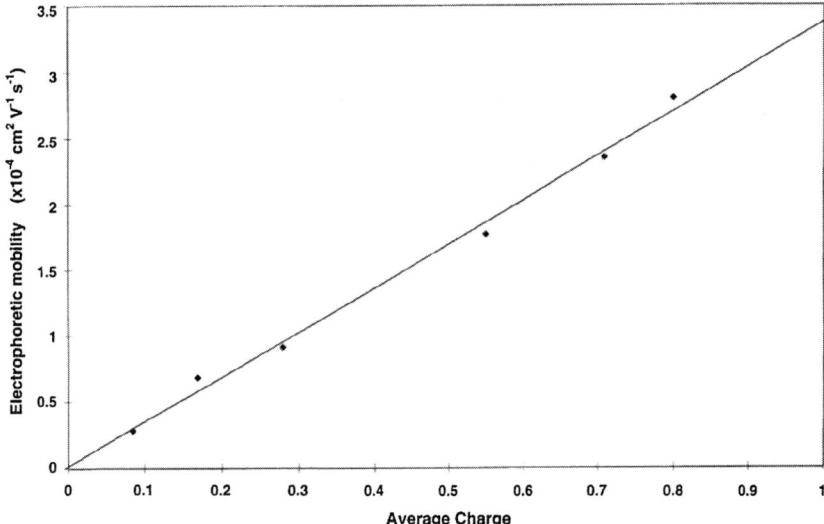

Figure 2.3. Graph showing measured electrophoretic mobility as a function of the calculated molecular charge for the peptide diglycine.

ferent pK_a values and a subsidiary equation derived which describes the charge difference between them. It can be shown that the charge difference between the two species will be maximised at a pH equivalent to the average of the two pK_a values [9].

2.3.3 Molecular Shape

It is often possible to separate isomeric species by free solution CE. Separation is even possible under conditions where the isomers carry the same charge. Figure 2.4 shows the separation of the methyl pyridines and ethyl pyridines in a lithium phosphate buffer at pH 2.5. In both cases

the 3 and 4 derivatives have a higher electrophoretic mobilities than the 2 derivatives. In addition the 3 and 4-ethyl-pyridines are also partially separated.

At a naive level these mobility differences can be explained by the 'shape' of the molecule. The protonated alkyl pyridine could be expected to be aligned in the electrical field with the positively charged end pointing to the cathode. This molecular alignment results in the 3 and 4 derivatives having a more streamlined shape than the 2 derivatives, and also in hindered rotations about two of the axes. This simple argument can also be applied for the dimethyl pyridines analysed under the same conditions. Figure 2.5 shows 2,6-dimethyl-pyridine to have the lowest elec-

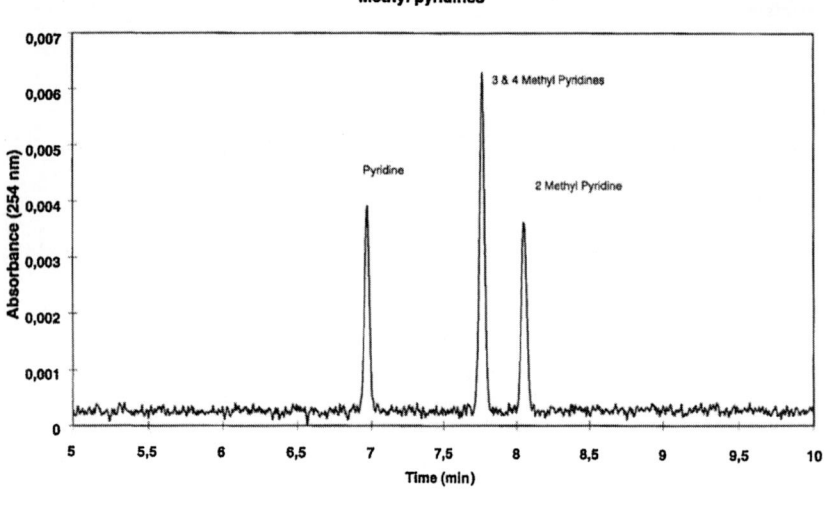

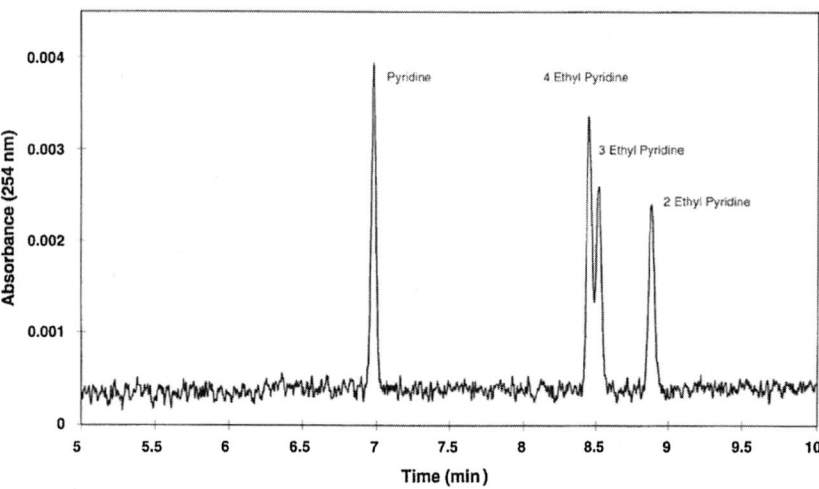

Figure 2.4. Electropherograms showing the separations of the methyl pyridines and the ethyl pyridines at pH 2.5.

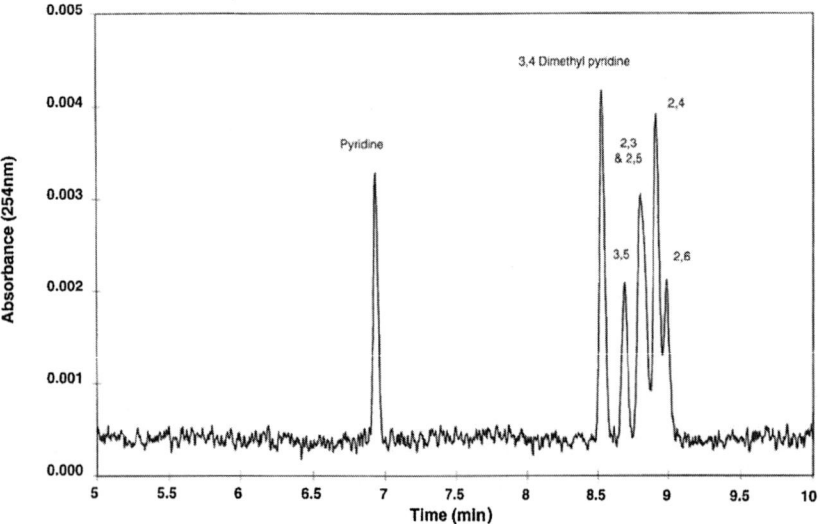

Figure 2.5. Electropherogram showing the separation of the dimethyl pyridines at pH 2.5.

trophoretic mobility and 3,4-dimethylpyridine to have the highest. The hindered

rotation model was formalised and combined with a size model to try and corre-

late the mobilities of a range of alkyl pyridines [10]. The alkyl pyridines were assumed to be aligned in the electrical field about an axis running through the centres of mass and charge. The model employed three parameters: the overall Van der Waals radius, the distance between the centres of mass and charge, and the radius of the aligned molecule. The three term model was shown to provide a good correlation with the measured electrophoretic mobilities of 21 alkyl pyridines.

2.3.4 Buffer Type and Concentration

The electrophoretic mobility of an analyte is also a function of the concentration and type of the buffer salts employed. The electrophoretic mobilities of the oligoglycines was shown to depend on the concentration of the buffer [11]. Issaq and others found that the electrophoretic mobility of dansylalanine was proportional to the inverse of the square root of the buffer concentration [12]. The use of phosphate and acetate buffers gave similar linear relationships between electrophoretic mobility and buffer concentration but with different gradients.

The electrophoretic mobilities of some oligoalanines and oligoglycines were measured using different buffers, buffer concentrations, and low and high pHs [13]. These peptide oligomers are useful model analytes as the amino and carboxylic acid termini mean that both the anionic and cationic forms of the peptides could be studied according to the buffer pH used. The anionic forms were studied using high pH phosphate and borate buffers, and the cationic forms using low pH phosphate and citrate buffers. Electrophoretic mobilities were determined using a range of buffer concentrations. The electrophoretic mobilities were shown to be a function of the ionic strength of the buffer rather than the simple concentration. The ionic strength forms the basis for comparing the effects of different buffer types. The measured electrophoretic mobilities of the oligoglycines and oligoalanines under the different conditions were related to common limiting mobilities (zero buffer concentration). The approach involves treating deviations from ideality in ionic transport of the analytes in the same way as deviations from ideality in ionic equilibria and follows the approach of Stokes and Robinson [13]. At 298 K the measured mobility is related to the limiting mobility and the ionic strength by equation (2.7).

$$\mu = \mu_0 - \frac{(0.229\,\mu_0 + 3.12 \times 10^{-8})I^{0.5}}{1 + 3.28\,aI^{0.5}}$$

$$(2.7)$$

Where μ is the measured electrophoretic mobility, μ_0 the limiting mobility, a the sum of the analyte and counter ion radii, and I the ionic strength of the buffer. The ionic strength is defined in equation (2.8).

$$I = 0.5 \sum_j m_j z_j^2 \qquad (2.8)$$

Where m_j is the molality of the buffer ion and z_j the charge.

The use of equations (2.7) and (2.8) meant that the data obtained in the different buffer systems and at different buffer concentrations could be compared. By using equations (2.7) and (2.8) the same values for the limiting mobilities (μ_0) were obtained from the different buffers and for both the positively and negatively charged forms of the analytes.

The above discussion showed how the electrophoretic mobilities of achiral molecules can be related to their size, charge, and shape. These principles are also important in considering the separation of enantiomers. In CE the separation of enantiomers depends upon the transient or permanent formation of diastereoisomeric species. Permanent diastereoisomers may have sufficiently different shapes or pK_a values. Electrophoretic mobility differences can arise from the shape differences or by inducing a charge difference via choosing a pH in between the two pK_a values. For transient species, such as the inclusion complex formed between a charged analyte and a neutral cyclodextrin, all three factors may be important. If the two enantiomers have different affinities for the cyclodextrin then the effective sizes and hence electrophoretic mobilities will be different. If the two enantiomers are orientated differently in the cyclodextrin cavity then this may lead to either a difference in the effective shape or may alter the effective pK_a values of the analyte to different extents.

2.4 Electroosmotic Mobility

2.4.1 Background Theory

Electroosmosis is the bulk flow of the buffer through the capillary under the influ-

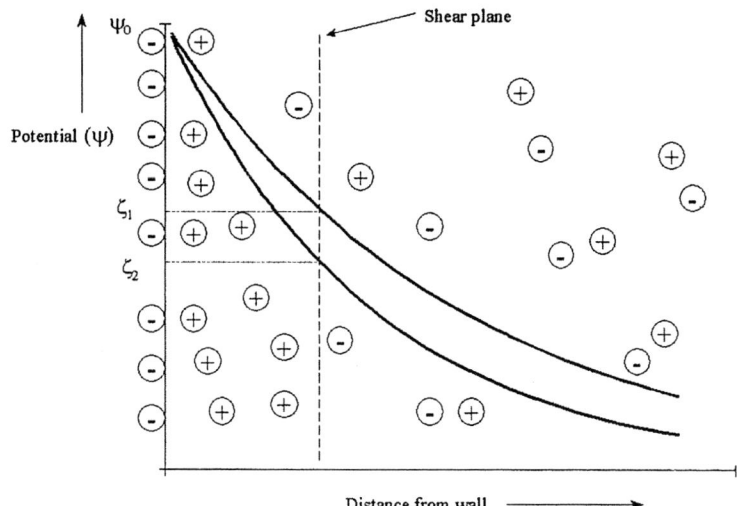

Figure 2.6. The electrical double layer at the capillary wall-buffer solution interface.

ence of the applied electrical field. As the electroosmotic mobility is the same for all analytes it does not lead directly to separation but does alter the extent to which separative processes can be effective. Electroosmosis arises as a result of the potential difference between the capillary wall and the bulk buffer solution. A full theoretical description of electroosmosis is very complex and is beyond the scope of this book and the following discussion contains many simplifications and approximations. For a detailed treatment of electroosmosis the reader is referred to specialised texts such as [14].

Charges on the capillary wall arise from either adsorbed ions or by ionisation of the chemical groups on the surface of the wall. Most CE is carried out in fused silica capillaries and so the state of the silanol groups on the surface is important. Ionisation of the silanol groups leads to a negative charge at the capillary surface. The negative charge on the silica surface attracts cations from the buffer solution and repels anionic ones. This arrangement of negative and positive charges at the capillary-buffer interface is called the electrical double layer and is illustrated in Figure 2.6.

The charge separation at the interface gives rise to a potential difference between the surface of the capillary and the buffer solution. The potential difference is greatest at the surface but decreases with distance into the buffer where there is less separation of charge. The potential drops with distance in an exponential manner. The rate at which the potential drops also

depends upon the ionic strength of the buffer solution as described in equation (2.9) (p 25 in [14]).

$$\psi = \psi_0 e^{-\kappa x} \qquad (2.9)$$

Where x is the distance from the surface, ψ_0 is the potential at the surface, ψ is the potential at x and κ is the Debye-Hückel parameter.

The Debye-Hückel parameter, κ, is proportional to the square root of the ionic strength and so it follows that the drop in potential will be greater for more concentrated buffers. The quantity $1/\kappa$ is also called the double layer thickness. The potential at the surface is determined by the charge on the capillary wall. With fused silica the charge on the capillary wall depends upon the charge on the silanols which is controlled by the pH of the buffer solution. At low pHs the silanols are all fully protonated and so the charge is low whilst at alkaline pHs the silanols will all be charged.

The cations next to the capillary wall will be strongly bound by electrostatic forces but those slightly further out into the bulk buffer solution are more mobile. When an electrical field is applied along the capillary the mobile cations near to the wall are drawn to the cathode and tend to draw the rest of the buffer solution with them. This bulk flow of the buffer solution is called electroosmosis.

Because the cations right next the capillary wall are stationary and those slightly further away are drawn towards the cathode there is a plane of shear close

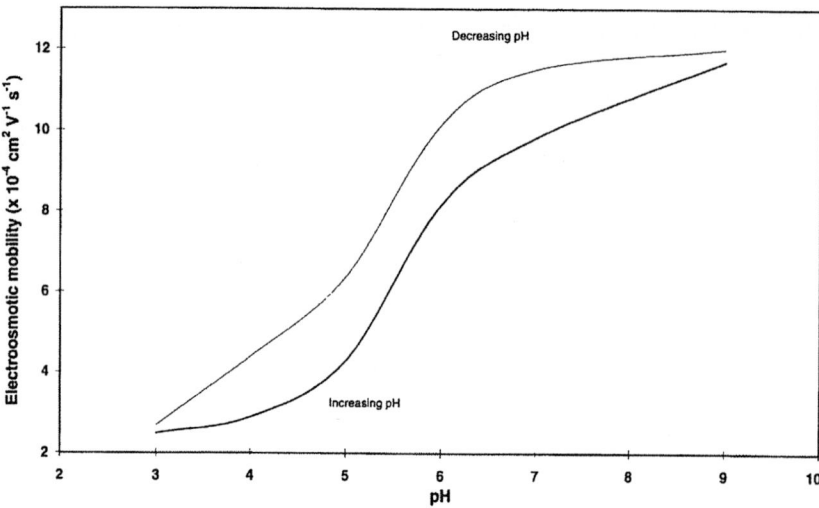

Figure 2.7. Electroosmotic mobility as a function of buffer pH.

to the surface. The potential at this plane of shear is called the ζ (zeta) potential. The ζ potential is important as the electroosmotic mobility is directly proportional to it as shown in equation (2.10).

$$\mu = \frac{-\varepsilon\,\varepsilon_0 \zeta}{\eta} \qquad (2.10)$$

Where ε_0 is the permittivity of a vacuum, ε is the dielectric constant, and η is the viscosity of the buffer.

Whilst the velocity at the plane of shear is zero it increases into the bulk buffer until a constant value is achieved far from the surface. The normal assumption in CE is that the distance over which the velocity varies is very small relative to the diameter of the capillary and so the velocity is effectively constant. One advantage of an electrically driven separation system over one driven by pressure is that the flow profile is much flatter leading to less dispersion of the analyte bands.

2.4.2 Experimental Conditions

In Section 2.4.1 the theory indicated that the size of the electroosmotic mobility would be expected to depend upon the the properties of the buffer such as the pH, ionic strength and the viscosity.

The influence of the buffer pH on the electroosmotic mobility can be determined by the injection of a neutral marker and measurement of the migration time. Lukacs and Jorgenson showed sigmoidal relationships with pH for capillaries made from silica, pyrex, and teflon [15]. A sigmoidal relationship was also shown by Kohr and Engelhardt using a fused silica capillary and 10 mM sodium phosphate

buffers at pHs between 3 and 9 [16]. Different values for the electroosmotic mobility were obtained according to whether the buffers were used in ascending or descending order of pH. The size of the hysteresis seen can be decreased by flushing the capillary with the buffer for longer periods of time. The hysteresis is shown in Figure 2.7 and demonstrates the need to equilibrate the capillary properly prior to analysis in order to obtain reproducible analysis conditions.

The shape of the curves in Figure 2.7 follows that which might be expected from the titration of the acidic silanol groups. Figure 2.7 also shows that whilst the electroosmotic mobility is stable at high and low pHs, small changes of pH in the neutral region will lead to relatively large changes in the electroosmotic mobility. This variability has significant implications for the reproducibility of CE methods operated at neutral pHs.

The size of the electroosmotic mobility depends upon the concentration of the buffer. The effect of varying buffer concentration on electroosmotic mobility was investigated using phosphate and borate buffers at pH 8.0, and the Good's buffers ACES, HEPES and HEPPSO at pH 7.5 [17]. The electroosmotic mobility was shown to decrease linearly with the log of the buffer concentration for each of the buffer systems. The different buffer systems gave rise to separate lines with different intercepts and gradients. It was shown that the data from the different buffers could be combined by plotting the electrophoretic mobility against the log of the ionic strength rather than concentration. All the data obtained at pH 7.5 fell on a

single straight line regardless of the type of buffer used and a similar result was obtained with the pH 8.0 data. These experimental results are therefore consistent with the expectations from the theory discussed above.

Altria and Simpson showed that the electroosmotic mobility could be reduced and sometimes reversed by increasing the concentration of different alkyl trimethylammonium bromides [18, 19]. The electroosmotic mobility was inversely proportional to the log of the concentration and the greatest change (and reversal) was seen with longer alkyl chain lengths. As high pH buffers are prepared by the addition of an alkali to an acid or salt, an increase in the buffer pH will also result in an increase in the ionic strength. Because increasing pH and ionic strength have the opposite effects on the electroosmotic mobility a higher pH buffer may on some occasions give a lower electroosmotic mobility [20].

Because of the influence of the surface charge many workers have investigated the use of coated capillaries to reduce or control the size of the electroosmotic mobility. Kohr and Engelhardt for example used two types acrylamide coating, one was uncharged and the other contained sulphonic acid groups [16]. The sulphonic acid coated capillary showed a much smaller change in electroosmotic mobility with pH than the fused silica capillary. The differences arise because sulphonic acid groups are very acidic and so will be charged right across the normal pH range for CE. The capillaries with the neutral acrylamide coating were also shown to give electroosmotic mobilities which varied with buffer pH. With the neutral coating the size and the pH dependence of the electrophoretic mobilities was much smaller than with fused silica. The coated capillaries also showed less hysteresis with pH changes and better reproducibility of electroosmosis.

The electroosmotic mobility arises because of the potential difference between the capillary wall and the buffer solution. One experimental approach to reduce and control the size of the electroosmotic mobility has been to apply an external radial field to counteract the potential difference. By applying different external fields Lee, Blanchard and Wu were able to reduce and then reverse the measured electroosmotic flow velocity [21]. In a variant of this approach Culbertson and Jorgenson used an external field which varied along the capillary [22]. By varying the external field during the run the size and sign of the net mobility could be varied and the analytes repeatedly passed back and forth through the detector.

In equation (2.5) the resolution was shown to be proportional to the square root of the applied voltage. As the analysis time is inversely related to the applied voltage the highest voltage possible should in principle be used. In practice however many workers have found that there is an optimum value for the applied voltage [12]. This optimum is thought to arise due to significant contributions to band broadening which can arise due to thermal gradients within the capillary [23]. Both electroosmotic and electrophoretic mobilities change with temperature via the viscosity of the buffer and so thermal gradients will lead to a range of analyte mobilities. The thermal gradients arise because of the heat generated within the capillary by the passage of a current across a potential difference.

CE may be carried out in solvents other than water and Valko and co-workers have measured the electroosmotic mobilities in a range of pure organic solvents without the presence of buffer salts [24]. The highest electroosmotic mobility was obtained in acetonitrile which gave a value of more than double that seen in water.

2.5 CE Instrumentation

2.5.1 Introduction

Section 2.5 is a short summary of the instrumentation used in CE and the importance and performance of the various components. Section 2.5 is intended as a brief overview and for more detailed information the reader is referred to the additional reading listed at the end of the chapter.

The basic instrumentation used in CE is relatively simple and a typical system is shown schematically in Figure 2.8.

The typical CE system consists of a high voltage power supply, an injection system, the separation capillary, a detector, and a data collection and processing system. Commercial CE systems are widely available and employed by most workers particularly those in commercial organisations. Most commercial systems employ an autosampler and computer control. Because of the relative simplicity of CE many modified and home built sys-

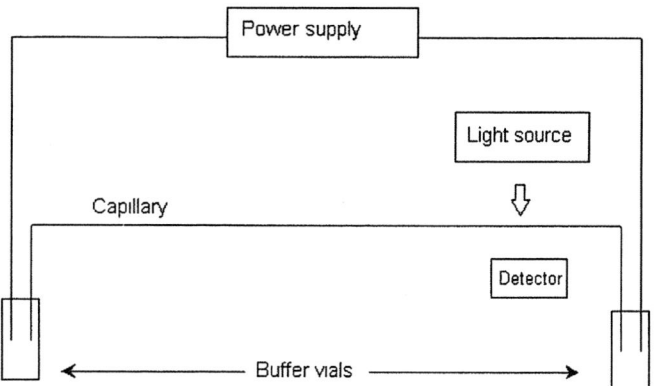

Figure 2.8. The basic CE system.

tems have been constructed by researchers interested in the principles of the technique or in instrument development.

2.5.2 The Power Supply

Power supplies capable of generating potential differences of up to 30 kV DC are commonly used. Usually the polarity of the power supply is switchable such that the detector end of the capillary can coincide with either the cathode or the anode. As the analysis time depends upon the voltage it is important that the power supply can give reproducible voltages. High voltages are desirable because of the higher efficiencies and shorter analysis times that can be obtained. Equation (2.4) shows that the analysis time is inversely proportional to voltage. Voltages of greater than 30 kV are less common because of the greater problems with electrical insulation and break down of the electrical field and the capillary. The use of voltages of up to 120 kV have been reported but, whilst they can lead to very high efficiencies and resolution, special construction procedures and precautions are required [25]. The voltages used in CE are high and whilst the typical currents (and hence power) are very low there is a risk of electrical shock if proper safety procedures are not followed.

2.5.3 The Capillary

The separation capillaries are normally made from fused silica with a polymer coating on the exterior surface to protect the fused silica from scratches. The interior surface may be coated with various polymers to either control or reduce the electroosmotic flow. Fused silica is popular because of its transparency to light, electrical insulation properties, flexibility and wide availability and low cost due to its use in GC. The capillary dimensions vary according to the application but are typically between 20 µm and 75 µm in internal diameter, 375 µm external diameter, and between 20 cm and 100 cm in length. Narrow capillaries are favoured for high efficiency separations but disfavoured for detection sensitivity. The electrical current and so the joule heating increases as the square of the diameter and so band broadening due to thermal gradients is a greater problem with wider capillaries. With most CE systems the maximum detection path length is the diameter of the capillary and so wider capillaries give greater sensitivity. Longer capillaries lead to longer migration times. By rearrangement of equation (2.4) it can be seen that when a constant voltage is used the analysis time is proportional to the product of the total capillary length and the length of the capillary from the inlet to the detector $(l \cdot L)$. For a system having a fixed length from the detector to the capillary outlet of 7 cm, increasing the length L from 20 cm to 50 cm gives an increase in the analysis time by a factor of 5.3. Longer capillaries do give a bigger volume and surface area for the dissipation of Joule heat and this can be important in the cases where high buffer concentrations and voltages are needed to maximise efficiency.

Most commercial systems employ either air or liquid cooling systems to remove the heat generated. Removal of heat is important as higher capillary temperatures lead to a reduction in electrical resistance and so even higher currents and electrical heating. Efficient heat removal means that higher voltages and so shorter analysis times can be employed. A useful tool to measure the system performance is the Ohm plot i.e. voltage applied vs measured current. If the heat generated is being effectively removed the current generated will be directly proportional to the applied voltage. Operation in a non linear part of the curve can lead to poor reproducibility of results.

2.5.4 The Injector

The function of the injector is to ensure that a small but reproducible amount of sample is injected into the capillary. Unless the length of the injected sample plug is small relative to the capillary length it can becomes a significant contributor to band broadening [25, 26]. As a general rule of thumb the length of injected plug should be less than 1% of the length of the capillary.

Injection systems in CE are simple and three approaches are commonly employed: hydrostatic, hydrodynamic, and electrokinetic. With hydrostatic injection the capillary is dipped into the sample vial which is then raised above the level of the outlet vial and the sample is loaded onto the column by siphoning. The amount injected depends upon the sample concentration, the time of injection, the height differential, the capillary dimensions and the viscosity.

With hydrodynamic injection the capillary is placed in the sample vial and a pressure differential applied across the capillary. The differential is achieved by either applying a small positive pressure at the inlet end or a reduced pressure at the outlet end. The amount of sample injected depends upon the sample concentration and the volume of the sample. The Poisseuille equation for flow velocity is used to determine the length of the sample plug injected by making the assumption that the flow rate is constant [27]. The volume injected is then determined from the length of the sample plug and the equation giving the volume of a cylinder. The volume injected is given by equation (2.11).

$$\text{Volume} = \frac{\Delta P \pi t d^4}{128 \eta L} \qquad (2.11)$$

Where ΔP is the pressure differential, t is the injection time, d is the capillary diameter, η is the sample viscosity, and L is the length of the capillary.

For a hydrostatic injection the pressure differential in equation (2.11) is deter-

mined from the product of the height differential, the gravitational constant g, and the density of the sample solution.

With electrokinetic injection the capillary is dipped into the sample and a voltage applied for a certain time. The amount injected depends upon the sample concentration, the size of the voltage, the time, the electrophoretic and electroosmotic mobilities, and the capillary dimensions.

The most common injection methods on commercial CE systems are hydrodynamic and electrokinetic. Hydrodynamic methods are preferred in that the sample injected is representative of the sample in the vial. With electrokinetic approaches there is a bias in favour of the sample components with a higher electrophoretic mobility. Electrokinetic methods are however the only ones possible when the capillary has very viscous contents e.g. capillaries with cross linked gels. The volume injected in CE is typically less reproducible than that obtained in HPLC and so quantitative analysis typically requires either an internal standard or an area percent method.

2.5.5 The Detector

The detector is normally placed as close to the capillary outlet as is practical as any separation which occurs after the detector is wasted. Equation (2.5) shows that the resolution is maximised as l approaches L.

Detection relies on the analyte of interest having some property which differentiates it from the background buffer. Many detection modes have been used in CE but only a few of them are available commercially. Commercial systems include UV absorbance, fluorescence, electrochemical, and mass spectrometry as detectors. UV absorbance is the most commonly used detection technique in CE because it offers the best compromise of universality, sensitivity, and simplicity. Most organic molecules have a UV chromaphore and UV detectors are simple, reliable, moderately sensitive, and give a linear response. Many CE UV absorbance detectors work by shining light directly across the diameter of the capillary. The size of the absorbance is given by the Beer-Lambert law and given in equation (2.12).

$$\text{Absorbance} = -\varepsilon c l_d \qquad (2.12)$$

Where ε is the molar extinction coefficient at the wavelength of interest, c is the

analyte concentration and l_d is the detector path length. The extinction coefficient is fixed and the analyte concentration is ideally limited to less than 1% of the buffer concentration by the need to avoid electrophoretic dispersion [28]. Concentration sensitivity in CE is usually inferior to HPLC as in most systems the detector path length l_d is limited to the capillary diameter. A longer path length may be obtained by the use of either a bubble flow cell within the capillary or a internal flow cell. With the internal flow cell, or so called Z-shaped flow cell, the flow path is bent through a right angle and the UV light shines along part of the capillary rather than across it. This approach can lead to a significant increase in sensitivity with only a small loss in efficiency [29].

The sensitivity of UV detectors has also been improved by the development of more stable light sources which reduce the background noise in the signal. As the path lengths are short and very intense light sources are employed, it is usually possible to operate at lower wavelengths than are routine in HPLC. For example detection wavelengths of 200 nm are common in CE.

The use of diode array and MS detectors provides information on the UV and mass spectra of the analytes and this aids the determination of the identities of the sample components.

Despite the significant improvements in recent years the exploitation of the full potential of CE is still limited by the need for more sensitive UV detectors.

Fluorescence provides much greater sensitivity than is possible by UV absorbence but is limited by the number of compounds which are naturally fluorescent. Many compounds can be tagged with fluorescent molecules but this adds to the complexity of the method and brings concern about different degrees of derivatisation for different analytes. In addition environmental factors can lead to different degrees of fluorescence quenching for different analytes. The fluorescence quenching that can be observed after the inclusion of an analyte into a cyclodextrin may well be significant for enantiomer separations.

2.5.6 The Data System

Most CE systems employ a data system to collect and process the signals from the detector. An important difference in signal processing from that used in HPLC is that the areas of the CE peaks are divided by their migration times to give corrected areas. In CE the slower moving analytes spend longer in the detector (by definition) and so for the same concentration and extinction coefficient give rise to larger areas. Division of the peak area by the migration time of the peak corrects for the effect of the different analyte velocities.

2.6 Conclusion

The equipment employed and the basic separation principles in CE are both relatively straightforward. Separation in CE requires differences in the electrophoretic mobilities of the analytes and these can be obtained via differences in their effective charge, size, or shape. The key to the separation of enantiomers in CE therefore lies in the introduction of additional agents which induce differences in their effective charges, sizes, or shapes.

References

Further Reading

Khaledi, M.G. (Ed.), High Performance Capillary Electrophoresis. Theory Techniques and Applications, J. Wiley, New York, **1997**.

Kok, W. Capillary Electrophoresis: Instrumentation and Operation, Vieweg Publishing, Wiesbaden, **2000**.

[1] Giddings, J.C. Unified separation science, John Wiley & Sons, New York, **1991**.

[2] Terabe, S.; Yashima, T.; Tanaka, N.; Araki, M. Separation of oxygen isotope benzoic acids by capillary zone electrophoresis based on isotope effects on the dissociation of the carboxyl group., *Anal. Chem.* **1988**, *60*, 1673–1677.

[3] Giddings, J.C. Generation of variance, "Theoretical Plates", Resolution and Peak Capacity in Electrophoresis and Sedimentation., *Sep. Sci.* **1969**, *4*, 181–189.

[4] Jorgenson, J.; Lukacs, K.D. Zone electrophoresis in open-tubular glass capillaries., *Anal. Chem.* **1981**, *53*, 1298–1302.

[5] Offord, R.E. Electrophoretic mobilities of peptides on paper and their use in the determination of amide groups. *Nature* **1966**, *211*, 591–593.

[6] Rowe, R.C.; Wren, S.A.C.; McKillop, A.G. Molecular size/shape effects in the separation of the monosubstituted alkylpyridines using capillary electrophoresis., *Electrophoresis* **1994**, *15*, 635–639.

[7] Survay, M.A.; Goodall, D.M.; Wren, S.A.C.; Rowe, R.C. Oligoglycines and oligoalanines as tests for modelling mobility of peptides in capillary electrophoresis., *J. Chromatogr.* **1993**, *636*, 81–86.

[8] Compton, B.J.; O' Grady, E.A. Role of Charge Suppression and Ionic Strength in Free Zone Electrophoresis of Proteins., *Anal. Chem.* **1991**, *63*, 2597–2602.

[9] Wren, S.A.C. Optimization of pH in the Electrophoretic Separation of 2-, 3-, and 4-Methylpyridines., *J. Microcol. Sep.* **1991**, *3*, 147–154.

[10] McKillop, A.G.; Smith, R.M.; Rowe, R.C.; Wren, S.A.C. Modelling and Prediction of Electrophoretic Mobilities in Capillary Electrophoresis: Separation of Alkylpyridines., *Anal. Chem.* **1999**, *71*, 497–503.

[11] Survay, M.A.; Goodall, D.M.; Wren, S.A.C. Rowe, R.C. Effects of pH and Buffer Concentration on the Separation of Oligoglycines by Capillary Electrophoresis., *Anal. Proc.* **1993**, *30*, 477–479.

[12] Issaq, H.J.; Atamna, I.Z.; Muschik, G.M.; Janini, G.M. The Effect of Electric Field Strength, Buffer Type and Concentration on Separation Parameters in Capillary Zone Electrophoresis., *Chromatographia* **1991**, *32*, 155–161.

[13] Survay, M.A.; Goodall, D.M.; Wren, S.A.C.; Rowe, R.C. Self-consistent framework for standardising mobilities in free solution capillary electrophoresis: applications to oligoglycines and oligoalanines., *J. Chromatogr. A* **1996**, *741*, 99–113.

[14] Hunter, R.J. Zeta potential in colloid science, Academic press, London, **1981**.

[15] Lukacs, K.D.; Jorgenson, J.W. Capillary Zone Electrophoresis: Effect of Physical parameters on Separation Efficiency and Quantitation., *J.H.R.C. & C.C.* **1985**, *8*, 407–411.

[16] Kohr, J.; Engelhardt, H. Capillary Electrophoresis with Surface Coated Capillaries, *J. Microcol. Sep.* **1991**, *3*, 491–495.

[17] Van Orman, B.B.; Liversidge, G.G.; McIntire, G.L.; Olefirowicz, T.M.; Ewing, A.G. Effects of Buffer Composition on Electroosmotic Flow in Capillary Electrophoresis., *J. Microcol. Sep.* **1990**, *2*, 176–180.

[18] Altria, K.D.; Simpson, C.F. High Voltage Capillary Zone Electrophoresis: Operating Parameters Effects on Electroendosmotic Flows and Electrophoretic Mobilities., *Chromatographia* **1987**, *24*, 527–532.

[19] Altria, K.D.; Simpson, C.F. The Effect of Electrolyte Chain Length on Electroendosmotic Flow in High Voltage Capillary Zone Electrophoresis, *Anal. Proc.* **1988**, *25*, 85.

[20] Vindevogel, J.; Sandra, P. Simultaneous pH and ionic strength effects and buffer selection in capillary electrophoretic techniques, *J. Chromatogr.* **1991**, *541*, 483–488.

[21] Lee, C.S.; Blanchard, W.C.; Wu, C.T. Direct Control of Electroosmosis in Capillary Zone Electrophoresis by Using an External Electric Field, *Anal. Chem.* **1990**, *62*, 1550–1552.

[22] Culbertson, C.T.; Jorgenson, J.W. Increasing the Resolving Power of Capillary Electrophoresis Through Electroosmotic Flow Control Using Radial Fields, *J. Microcol. Sep.* **1999**, *11*, 167–174.

[23] Foret, F.; Deml, M.; Bocek, P. Capillary zone electrophoresis. Quantitative study of the effects of some dispersive processes on the separation efficiency, *J. Chromatogr.* **1988**, *452*, 601–613.

[24] Valkó, I.E.; Sirén, H.; Riekkola, M. Characteristics of electroosmotic flow in capillary electrophoresis in water and in organic solvents without added ionic species., *J. Microcol. Sep.* **1999**, *11*, 199–208.

[25] Hutterer, K.M.; Jorgenson, J.W. Ultrahigh-Voltage Capillary Zone Electrophoresis., *Anal. Chem.* **1999**, *71*, 1293–1297.

[26] Huang, H.; Coleman, W.F.; Zare, R.N. Analysis of factors causing peak broadening in capillary zone electrophoresis, *J. Chromatogr.* **1989**, *480*, 95–110.

[27] Chien, R-L. in High Performance Capillary Electrophoresis, (ed. M. G. Khaledi), J. Wiley & Sons Inc, New York, **1998**.

[28] Mikkers, F.E.P.; Everaerts, F.M.; Verheggen, Th. P.E.M. Concentration distributions in free zone electrophoresis, *J. Chromatogr.* **1979**, *169*, 1–10.

[29] Chervet, J.P.; Van Soest, R.E.J.; Ursem, M. Z-Shaped flow cell for UV detection in capillary electrophoresis, *J. Chromatogr.* **1991**, *543*, 439–449.

3 Modelling Enantiomer Separation by CE

3.1 The Use of Models

The first step in understanding the enantiomeric separation process in CE is the development of a simple physical model. The purpose of the simple physical model is an adequate description of the underlying interactions between the chiral selector and the two enantiomeric forms of the analyte. The mathematical model will follow once the physical model has been established.

Before physical concepts are discussed it is important to ask the question: what is the purpose of the model? This question is an important one because it helps to define both the level of approximation that is appropriate, and the limitations of the final model.

An interesting aspect of models of many physical processes is the exploration of the acceptable compromise between accuracy and accessibility. The model must reflect the observed behaviour in the real world, but will be of limited value if only skilled theoreticians can understand what it means. The position of the compromise between accuracy and accessibility will vary according to the purpose to which the model is applied and can be regarded as a subjective decision. It is likely that several models, or variants of the same model differing in complexity, will be used in different circumstances.

A good example is that of the gas laws. Boyle's law is adequate for a general understanding of the operating principle of a hot air balloon, but is unlikely to be sufficient for a chemical engineer designing a complex gas phase reactor.

Analytical chemistry is a practical subject which generates data for other purposes. Models in analytical chemistry are rarely ends in themselves but should be aimed at helping the practitioners in understanding and applying the separation process. Several models have been developed to describe the separation of enantiomers in CE. These models differ in their complexity and accuracy and different ones will be appropriate in different circumstances.

3.2 Background

The simple equilibrium model was developed to try and explain trends observed both in the literature and in my own experience. As was detailed in Chapter 1 the area of enantiomer separation was one of great potential right from the early days of CE. The ability to produce highly efficient separations using simple chiral additives held out great promise. Several workers had shown very pleasing results from additives such as cyclodextrins.

A careful examination of the early data showed that in most cases the resolution was shown to increase with increasing chiral selector concentration. In most cases a linear relationship between resolution and chiral selector concentration was seen. Fanali and Bocek, for example, showed that the resolution of the enantiomers of tryptophan increased linearly with increasing concentration of α-cyclodextrin added to the separation buffer [1].

In other cases from the literature more complex patterns of behaviour were observed, with the selectivity or resolution depending both upon the concentrations of chiral selector and organic solvent in the buffer. The graphs showing selectivity (or resolution) as a function of the selector concentration were typically non linear and maxima were often observed. The addition of organic modifier to the buffer could either increase or decrease the resolution depending upon the analyte and buffer components.

Fanali [2] worked on the resolution of terbutaline and propranolol enantiomers using both β-cyclodextrin and dimethyl-β-cyclodextrin. With terbutaline interesting patterns were seen using both of the cyclodextrins. At low cyclodextrin concentrations the resolution increased with increasing cyclodextrin concentration. At higher cyclodextrin concentrations the resolution increase tailed off until a maximum value was reached. As the cyclodextrin concentration was increased further the resolution declined. It was found that maximum resolution between the terbutaline enantiomers occurred at a concentration of 5 mM dimethyl-β-cyclodextrin or 15 mM β-cyclodextrin. The size of the maximum resolution obtained with dimethyl-β-cyclodextrin was superior to that seen with β-cyclodextrin.

The use of methanol in buffer systems was also investigated by Fanali [2] who examined propranolol in a phosphate buffer system containing 40 mM β-cyclodextrin. In the absence of methanol there was no resolution between the propranolol enantiomers. Some resolution was seen with the addition of 10% methanol to the buffer, and this was improved further with 30% methanol.

Guttman et al. [3] worked with cyclodextrins incorporated in polyacrylamide gels to resolve the enantiomers of dansylated amino acids. It was shown that the selectivity (as measured by the relative mobilities of the enantiomers) varied in a non linear fashion. The selectivity increased linearly at first but then the increase tailed off towards a plateau value. The slope of the graph showing the selectivity change with concentration varied with different amino acids. The rates of change of selectivity with changing concentration were believed to reflect the different affinities for the cyclodextrin.

The organic solvent content of the buffer system is also important. Twelve dansylated amino acids were analysed using a capillary containing 100 mM β-cyclodextrin incorporated in polyacrylamide [3]. Two buffer systems were used: one containing Tris/borate and a second with the same buffer salts with the addition of 10% methanol. Three of the dansylated amino acids gave lower resolution in the buffer containing methanol and nine gave higher

0009-5893/00/02   24-18 $ 03.00/0   © 2001 Friedr. Vieweg & Sohn Verlagsgesellschaft mbH

resolution. Particularly striking was the improvement in the resolution between the enantiomers of the derivatised aromatic amino acids phenylalanine and tryptophan.

Enantiomer separation by cyclodextrin modified MEKC was investigated by Nishi, Fukuyama, and Terabe [4]. Buffer systems containing SDS and various cyclodextrins (CDs) were used to explore which parameters were important in the resolution of the enantiomers of various compounds. The compounds included: thiopental, pentobarbital, 2,2,2-trifluoro-1-(9-anthryl) ethanol (anthryl ethanol), and 2,2'-dihydroxy-1,1'-dinaphthyl (dinaphthyl). At pH 9.0 with 50 mM SDS, γ-CD showed selectivity for all of the compounds, α-CD showed no selectivity at all, and β-CD and derivatised β-CD showed selectivity for some of the enantiomeric pairs. The effect of adding different concentrations of another chiral selector into a system containing 0.05 M SDS and 30 mM γ-CD was also determined. In the first set of experiments d-camphor-10-sulphonate was added at concentrations of 0, 20, and 40 mM. For thiopental, pentobarbital, and anthryethanol the resolution increased as a function of the d-camphor-10-sulphonate concentration, whereas for dinaphthyl it decreased. The effect of varying the concentration of l-menthoxyacetic acid between 0, 30, 60, and 90 mM in the same SDS/γ-CD system was also measured. For thiopental and dinaphthyl the resolution increased with increasing l-menthoxyacetic acid concentration. For pentobarbital and anthrylethanol the situation was more complex. The resolution increased up until a concentration of 60 mM l-menthoxyacetic acid and then declined. These trends were very interesting although it was acknowledged that the resolution could also be influenced by concentration because of changes to the micellar capacity factor and joule heating.

The methanol content of the buffer was also critical in the resolution of enantiomers of biphenanthrene dihydroxide using sodium cholate as the chiral selector. Cole et al. [5] showed that the resolution improved as the methanol content was increased from 0% to 6% and then 12%, but then declined as the methanol content was increased further to 18%.

Additional information on trends came from work on a commercial development compound, ICI 185,282. Other company workers had shown that by using various substituted β-CDs, the ^{19}F

NMR signals due to the enantiomers of ICI 185,282 could be split [6]. The size of the signal splitting increased with increasing methyl-β-cyclodextrin concentration. It was therefore decided to examine the use of methyl-β-cyclodextrin as a chiral buffer additive for the separation of ICI 185,282 enantiomers in CE. Due to the poor aqueous solubility of the compound 30% methanol had to be added to the buffer which contained 40 mM lithium phosphate at pH 6.0. It was found that whilst 2 mM of methyl-β-cyclodextrin gave baseline resolution of the two enantiomers, increasing the concentration to 4 mM lead to a decrease in resolution.

3.3 The Physical Model

As explained in the previous section, an examination of the experimental data showed interesting trends in the relationship between chiral selector concentration and the resolution between the enantiomers. Of particular interest was the observation that the resolution did not always increase linearly with chiral selector concentration, but often reached a maximum value and then declined. The existence of an optimum concentration is intriguing as it suggests that a competition process is in operation. It is clear that for any chiral discrimination to take place the enantiomers would have to interact in some way with the chiral selector. It is also obvious that the two enantiomers would either have to interact to different extents with the selector, or interact in a way which produced complexes which behaved in different ways. Any model which attempts to describe enantiomer separation in CE should reflect the interaction between the enantiomers and the chiral selector, and the importance of the chiral selector concentration.

The simple basic idea was that the enantiomers were involved in equilibrium reactions with the chiral selectors to produce transient complexes. The interaction of the β-blocker propranolol with β-cyclodextrin was chosen as a model system, but it was hoped that the model would have broad applicability.

If a charged enantiomer were to interact with a neutral chiral selector then the transient complex would have a lower charge/mass ratio and hence a lower electrophoretic mobility. It was expected that the equilibrium reaction involved would be very rapid and that this would give a

single observable peak which reflected average behaviour. The assumption of a rapid equilibrium seemed reasonable as the literature indicated that the sharpness of the peaks in the electropherograms was not altered by the presence of the chiral selector [4].

The belief in rapid equilibrium was supported by ^{1}H NMR data obtained from propranolol in aqueous solution, in both the presence and absence of β-cyclodextrin [7]. The results showed that whilst the propranolol signals shifted with increasing β-cyclodextrin they did not become significantly broader. Nor were separate proton signals observed for both the free and the complexed forms of propranolol. These results indicated that the exchange of propranolol between the free and complexed states was very rapid. Any equilibrium reaction which is fast on an NMR timescale is also expected to be fast on a CE timescale.

As the chiral selector concentration was increased the enantiomers would on average spend more time in the complexed rather than the free forms. The observed electrophoretic mobilities of the analytes would be a reflection of the proportion of time they existed in the complexed form, and the proportion they existed in the free form. At infinite selector concentration the observed electrophoretic mobility would approach the limiting value of that of pure complex.

If the enantiomers had different affinities for the chiral selector then, for the same selector concentration, the two enantiomers would spend different amounts of time in the complexed form. The different times would be reflected in different observed electrophoretic mobilities and hence the enantiomers could be separated. As a first order approximation the two enantiomer-chiral selector complexes can be thought of as having the same electrophoretic mobilities. As a general rule electrophoretic mobility is dominated by the size and charge of the analyte with small shape differences being much less important.

In this way the electrophoretic mobility difference will be zero at both zero and infinite chiral selector concentrations. It follows that there will be some intermediate chiral selector concentration which maximises the electrophoretic mobility difference the enantiomers. It is this optimum chiral selector concentration which is of most interest.

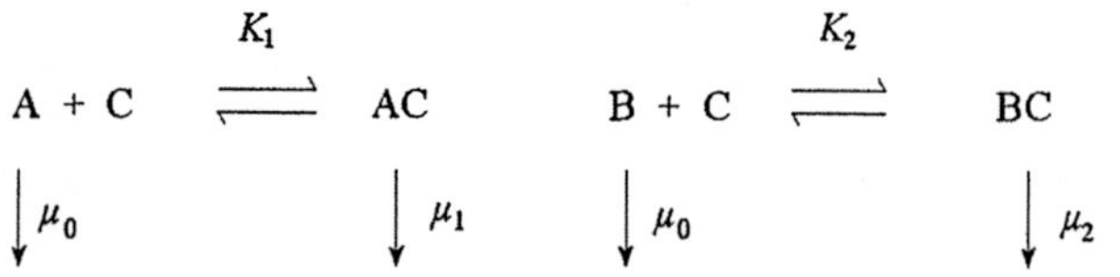

$$A + C \underset{}{\overset{K_1}{\rightleftharpoons}} AC \qquad B + C \underset{}{\overset{K_2}{\rightleftharpoons}} BC$$

$$\downarrow \mu_0 \qquad\qquad \downarrow \mu_1 \qquad \downarrow \mu_0 \qquad\qquad\qquad \downarrow \mu_2$$

Figure 3.1. The equilibria between the enantiomers **A** and **B** and the chiral selector **C**.

3.4 A Basic Mathematical Model

3.4.1 Electrophoretic Mobility

The physical case described above covers the situation of charged analyte enantiomers interacting rapidly with a cyclodextrin to form transient 1:1 complexes with the same limiting mobilities. Equal limiting mobilities for both complexes is obviously a simplifying assumption which cannot be expected to hold for all cases, but it is a useful first step. More complex situations will be covered in later models.

One of the normal aims in the development of routine analytical separation methods is the achievement of baseline resolution or greater between the two enantiomers. A resolution of baseline or greater is important if accurate estimations of the levels of the enantiomers are required. This is especially the case when one enantiomer is present at a much greater level than the other for example with optically pure pharmaceuticals. If the two enantiomers overlap to any significant extent then quantification becomes much more difficult.

As was discussed in Chapters 1 and 2 the resolution between two species has been shown to be a function of a number of parameters which relate to both the electrophoresis system and the analytes in question. Equation (3.1) was derived by Jorgenson and Lukacs [8] based upon the work of Giddings [9] and has been used by a number of others such as Terabe and coworkers [10].

$$R_S = \left(\frac{V}{32\,D}\right)^{0.5} \cdot \left(\frac{l}{L}\right)^{0.5} \cdot \frac{\Delta\mu_{ep}}{(\mu_{ep} + \mu_{eo})^{0.5}}$$
(3.1)

Where V is the voltage applied across both ends of the capillary; D is the diffusion coefficient of the analyte; L is the total length of the capillary; l is the length of the capillary form the inlet to the detector; $\Delta\mu_{ep}$ is the electrophoretic mobility difference between the two enantiomers; μ_{ep} is the mean electrophoretic mobility of the two enantiomers; and μ_{eo} is the electroosmotic mobility.

Equation (3.1) is derived from the simplest description of peak width, i.e. that in which band broadening arises only from molecular diffusion along the capillary axis. There are of course many other causes of band broadening such as: sample overloading; the contributions of the detection and injection lengths, and sample absorption on the capillary walls. These additional contributions were considered in Chapter 1 and will not be mentioned further at this point as they will merely cloud the issue. From equation (3.1) we can see that the applied separation voltage and capillary lengths play an important role in determining resolution and these were considered in detail in Chapters 1 and 2. In the next few pages we shall only consider the factors which are governed by the chemistry of the separation system i.e. the terms $\Delta\mu_{ep}$, μ_{ep}, and μ_{eo}.

From equation (3.1) we can see that to ensure good resolution we need to maximise the size of the electrophoretic mobility difference, $\Delta\mu_{ep}$, whilst also minimising the size of the term $(\mu_{ep} + \mu_{eo})^{0.5}$.

In the first instance we will consider the electrophoretic mobility difference between the two enantiomeric forms, $\Delta\mu_{ep}$. In order to describe $\Delta\mu_{ep}$ we must first derive expressions which describe the electrophoretic mobilities of the individual enantiomers [11].

The mobilities of the individual enantiomers are derived from the simple equilibrium situation which is shown in Figure 3.1.

In the absence of any chiral selector both enantiomers have the electrophoretic mobility μ_0. Enantiomer A interacts with chiral selector C to form the transient equilibrium complex AC which has the electrophoretic mobility μ_1. The position of the equilibrium is described by the equilibrium constant K_1. As the exchange between the free and complexed forms of enantiomer A is very rapid these two forms of enantiomer A will not be observed separately. Instead a third form of A will be

seen those behaviour is a composite of that of the free and complexed forms of A. The observed or apparent mobility of enantiomer A in the presence of the chiral selector, C, will depend on the proportion of time it is present in the free form, and the proportion of time it is present in the complexed form. The proportion of the time in each form, and hence the weighting of the two electrophoretic mobilities, μ_0 and μ_1, will depend upon the mole fractions. The total concentration of enantiomer A is equal to [A] + [AC] and so the proportion of the time that enantiomer A exists in the uncomplexed form will be equal to the fraction [A]/([A] + [AC]). Likewise the proportion of the time that that enantiomer A exists in the complexed form will be equal to the fraction [AC]/([A] + [AC]).

The apparent or observed electrophoretic mobility of enantiomer A will be described by the mole fraction weightings of μ_0 and μ_1 as given in equation (3.2).

$$\mu_{apparent} = \frac{\mu_0[A]}{[A] + [AC]} + \frac{\mu_1[AC]}{[A] + [AC]}$$
(3.2)

As [AC] = $K_1 \cdot$ [A] $\cdot$ [C] equation (3.2) can be simplified to equation (3.3).

$$\mu_{apparent} = \frac{\mu_0 + \mu_1 K_1[C]}{1 + K_1[C]}$$
(3.3)

Simple equilibrium equations of the type given in equation (3.3) are not new in separation science, it was found that they had already been used by Guttman et al. in chiral CE [3], and by Terabe and co workers to describe the modification of mobility in other equilibrium situations [12].

It was decided to investigate the consequences of equation (3.3) by computer modelling using suitable values of μ_0, μ_1, K_1 and [C]. Computer modelling can be very valuable as graphical information can be easier to consider than mathematical equations. Once the relevant spread sheets have been set up it is easy to explore the consequences of changes to all of the parameters used.

In Figure 3.2 the apparent electrophoretic mobility as a function of the chiral selector concentration is shown using the following parameters: $\mu_0 = 2 \times 10^{-4}$ cm^2 V^{-1} s^{-1}; $\mu_1 = 1 \times 10^{-4}$ cm^2 V^{-1} s^{-1}, K_1 = 100 M^{-1}; and the concentration range 0–0.1 M. The value for μ_0 is that which might be reasonably expected for a small positively charged organic molecule and follows from the examples and discussion in Chapters 1 and 2. The value for μ_1 is smaller than for μ_0 and represents the lower electrophoretic mobility which might be expected from a complex formed between a positive analyte and a neutral chiral selector such as a cyclodextrin. The value for K_1 represents an average typical value for an equilibrium reaction between a small organic molecule and a selector such as a cyclodextrin. The concentration range is that which might typically be obtainable with a chiral selector having good solubility.

Figure 3.2 shows that the apparent mobility when [C] = 0 is just μ_0. At high concentrations of the chiral selector the apparent mobility is starting to level off at a constant value. At high concentrations of C the $\mu_1 \cdot K_1 \cdot$ [C] term in the numerator will become very large in comparison to the μ_0 term, and the term $K_1 \cdot$ [C] will be very much larger than 1. So we can see that as [C] tends to infinity then the apparent mobility will tend to the limiting value of μ_1. This is useful as fitting actual data to the model can provides us with a way of estimating values for the limiting mobility.

In Figure 3.3 the apparent mobility curves obtained by using the equilibrium constants 10, 100, and 1000 M^{-1} are shown.

In Figure 3.3 the concentration range and values for μ_0 and μ_1 are the same as those used in Figure 3.2. The change in the equilibrium constants results in a change in the steepness of the apparent mobility curves. The larger the value of the equilibrium constant the more rapidly the apparent mobility tends to the limiting value of μ_1. We can see that for analytes which have a high affinity for the chiral selector the apparent electrophoretic mobility will be very sensitive to the addition of small concentrations of chiral selector. Conversely if the affinity of the analyte for the chiral selector is low then the addition of small concentrations of chiral selector will have little impact upon the apparent electrophoretic mobility of the analyte.

At this point it seems sensible to have a look at some experimental data to see if it is consistent with the proposed model. Figure 3.4 shows the change in the measured apparent electrophoretic mobility of the (S)-(–) enantiomer of the β-blocker

▷

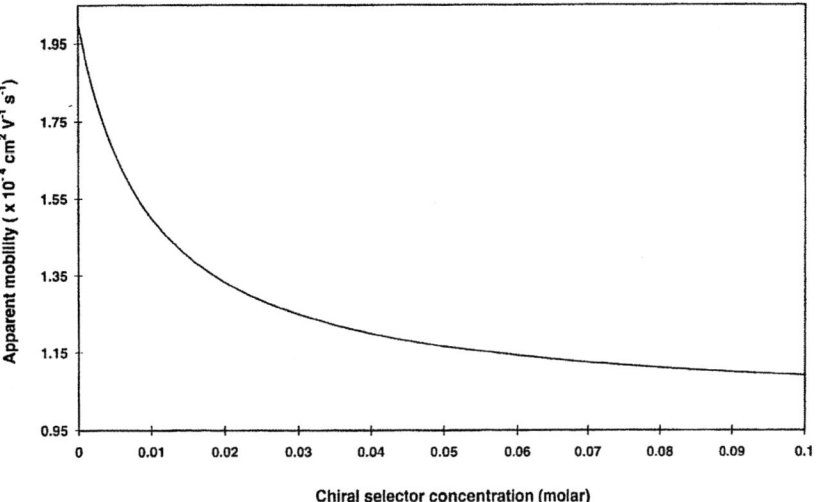

Figure 3.2. Apparent electrophoretic mobility (μ_a) as a function of [C] using $\mu_0 = 2 \times 10^{-4}$ cm^2 V^{-1} s^{-1} and $\mu_1 = 1 \times 10^{-4}$ cm^2 V^{-1} s^{-1} with $K_1 = 100$ M^{-1} using the selector range 0–0.1 M.

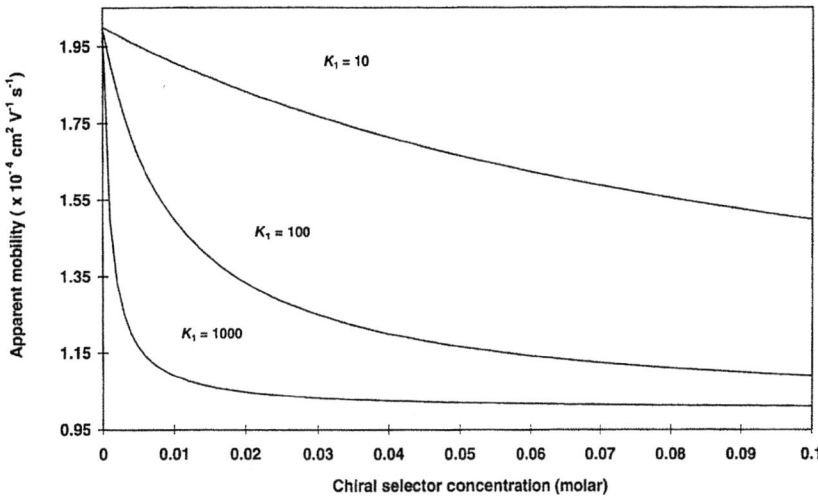

Figure 3.3. Apparent electrophoretic mobility (μ_a) as a function of [C] using $\mu_0 = 2 \times 10^{-4}$ cm^2 V^{-1} s^{-1} and $\mu_1 = 1 \times 10^{-4}$ cm^2 V^{-1} s^{-1} with $K_1 = 10$ M^{-1}, $K_1 = 100$ M^{-1}, and $K_1 = 1000$ M^{-1} using the selector range 0–0.1 M.

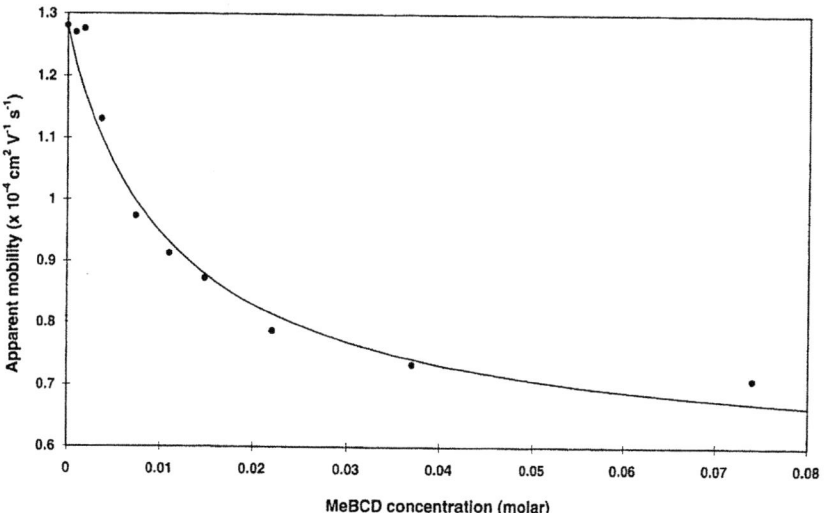

Figure 3.4. Measured apparent electrophoretic mobility for (S)-(–) propranolol as a function of the concentration of DMBCD.

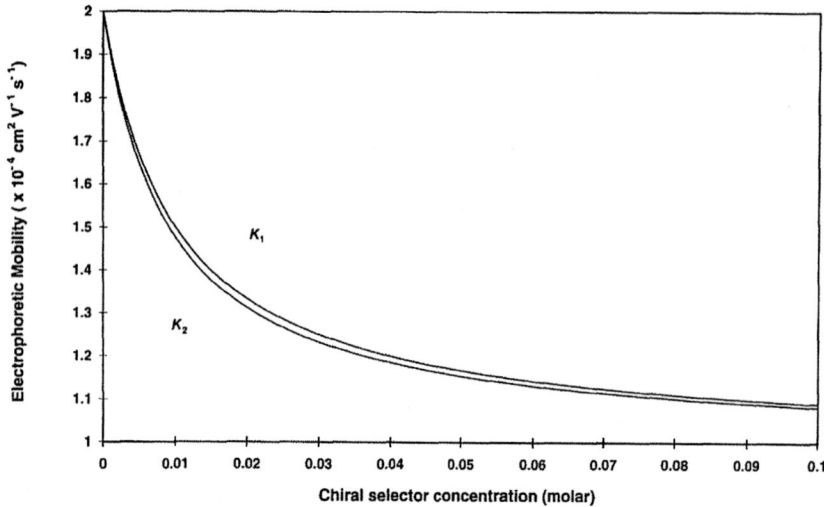

Figure 3.5. Apparent electrophoretic mobility of the enantiomers A and B using $\mu_0 = 2 \times 10^{-4}$ cm^2 V^{-1} s^{-1} and $\mu_1 = 1 \times 10^{-4}$ cm^2 V^{-1} s^{-1} with $K_1 = 100$ M^{-1} and $K_2 = 110$ M^{-1} using the selector range 0–0.1 M.

propranolol, with increasing concentrations of the chiral selector dimethyl-β-cyclodextrin (DMBCD).

The DMBCD is prepared by partially replacing the 2-, 3-, and 6-hydroxy groups of β-cyclodextrin with methoxy ones. Figure 3.4 shows a rapid drop in the apparent mobility of the (S)-(–) enantiomer with increasing DMBCD concentration. The curve has the same general form as those seen in Figure 3.3 and so these data are supportive of the model.

3.4.2 Electrophoretic Mobility Difference

In equation (3.1) we saw that the degree of resolution is influenced by: $\Delta\mu_{ep}$, μ_{ep}, and μ_{eo}. These mobility terms are determined both by the nature of the analyte and the chemistry of the separation buffer used. At this point we shall consider the term μ_{ep}, which determines the separation of the enantiomers and so is a key factor. Of course our ultimate aim is to model resolution but this is more complex mathematically than separation. We shall start with separation as the author prefers to try and break down a difficult problem into simpler parts.

In the discussion above we saw that the apparent electrophoretic mobility of enantiomer A could be described by equation (3.3). As $\Delta\mu_{ep}$ is the difference in the apparent electrophoretic mobility of the two enantiomers, $\Delta\mu_{ep} = \mu_{a1} - \mu_{a2}$. The apparent electrophoretic mobility of enantiomer B can be described by a similar

equation to (3.3), the only difference is that K_1 is replaced by K_2 (as a first approximation $\mu_2 = \mu_1$). The difference in the apparent electrophoretic mobilities of the two enantiomers is given by equation (3.4).

$$\Delta\mu_{ep} = \frac{\mu_0 + \mu_1 K_1 [C]}{1 + K_1 [C]} - \frac{\mu_0 + \mu_1 K_2 [C]}{1 + K_2 [C]}$$

(3.4)

Equation (3.4) can be rearranged to give equation (3.5) which is a more helpful form.

$$\Delta\mu_{ep} = \frac{[C](\mu_0 - \mu_1)(K_2 - K_1)}{1 + [C](K_1 + K_2) + K_1 K_2 [C]^2}$$

(3.5)

Equation (3.5) is a very important result so we shall take a little time to consider it in some detail. It is clear that the electrophoretic mobility difference, and hence the separation between the two enantiomers, will be strongly dependant upon both the nature of the analyte and the chiral selector that we choose. The concentration of the chiral selector will have a profound influence upon the electrophoretic mobility difference. Because of the [C] term in the numerator the electrophoretic mobility difference will be zero at zero chiral selector concentration i.e. no enantiomer separation in the absence of a chiral selector. What is less obvious is that [C]^2 term in denominator means that as [C] tends to infinity the electrophoretic mobility difference will tend to zero. If the electrophoretic mobility difference is zero both at zero and infinite chiral selector

concentration, the implication is that some intermediate selector concentration will maximise the electrophoretic mobility difference. The chiral selector concentration which will maximise the apparent electrophoretic mobility difference between the enantiomers is obviously of great practical interest.

The $\mu_0 - \mu_1$ term in the numerator of equation (3.5) means that $\Delta\mu_{ep}$ (the apparent mobility difference between the enantiomers) will increase as the mobility difference between the free and bound forms of the analyte increases. Thus a consideration of the mathematics alone lead to the early prediction [11] that chiral selectors carrying a charge opposite to that on the analyte would be useful. The $\mu_0 - \mu_1$ term also shows that neutral analytes can only be resolved using either a charged chiral selector or by using a charged achiral complexing agent in addition to the chiral selector. Equation (3.5) shows that if complexation with the chiral selector does not alter the electrophoretic mobility of the analyte there will be no separation.

Equation (3.5) shows that there will be no separation if both enantiomers have the same affinity for the chiral selector. We can also see that as the difference between the affinities of the two enantiomers for the chiral selector increases so will $\Delta\mu_{ep}$. The influence of the equilibrium constants K_1 and K_2 is a little complex as there are terms involving them in both the numerator and denominator of equation (3.5). Because of the complex nature of the influence of K_1 and K_2 it will be easier to understand them via the graphical representation shown in Figure 3.5.

Figure 3.5 is an extension of Figure 3.2 and also shows the apparent electrophoretic mobility of a second enantiomer whose affinity for the chiral selector differs from that of the first by a factor of 10% i.e. $K_1 = 100$, and $K_2 = 110$. The other parameters and the chiral selector concentration range are the same as used in Figure 3.2.

Because the second enantiomer has a greater affinity for the chiral selector than that of the first enantiomer the apparent electrophoretic mobility of the second enantiomer drops at a faster rate than that of the first. It is this greater decrease in apparent electrophoretic mobility that this leads to a difference in electrophoretic mobilities shown in Figure 3.5. At higher concentrations of the chiral selector the apparent mobilities converge and so the mobility difference becomes smaller.

The result is clearer if we plot, $\Delta\mu_{ep}$, the difference in the apparent electrophoretic mobilities of the two enantiomers (Figure 3.6).

From Figure 3.6 we can see that the difference in apparent electrophoretic mobility climbs steeply as the chiral selector concentration is increased from zero. At higher chiral selector concentrations the change in the apparent electrophoretic mobility difference begins to level off and reaches a maximum value. In Figure 3.6 the mobility difference is maximised at a chiral selector concentration of around 0.01 Molar. At higher concentrations the apparent electrophoretic mobility difference decreases although at a much slower rate than the initial increase. The result illustrated in Figure 3.6 is important because it reiterates the main consequences of equation (3.5), i.e. there will be an optimum concentration of chiral selector. Increasing the chiral selector concentration will not always lead to improved enantiomeric resolution and indeed can make it worse.

The influence of the equilibrium constants K_1 and K_2 can be determined by substituting different values for these parameters into equation (3.5).

In Figure 3.7 the parameters $\mu_0 = 2 \times 10^{-4}$ cm^2 V^{-1} s^{-1} ; $\mu_1 = 1 \times 10^{-4}$ cm^2 V^{-1} s^{-1}, $K_1 = 100$ M^{-1}; and the concentration range 0–0.1 M are used. Three values of K_2 are used giving values that are 1%, 5%, and 10% greater than that of K_1. The three curves show the differences in apparent electrophoretic mobility have the same general shape with similar optimum chiral selector concentrators. The difference between the curves in Figure 3.7 is in the size of the apparent electrophoretic mobility difference at the optimum chiral selector concentration. A doubling of the % difference between K_1 and K_2 leads approximately to a doubling of the mobility difference. From Figure 3.6 it is clear that for good resolution between the enantiomers there must be a significant difference in the relative affinities of them for the chiral selector. Big differences in relative affinities depends upon the correct choice of chiral selector as will be seen later.

The absolute size of the equilibrium constants is also important and will be investigated now. Figure 3.8 shows the apparent mobility differences which arise from three sets of equilibrium constant which differ in their absolute values, but not in the relative affinities of the two enantiomers.

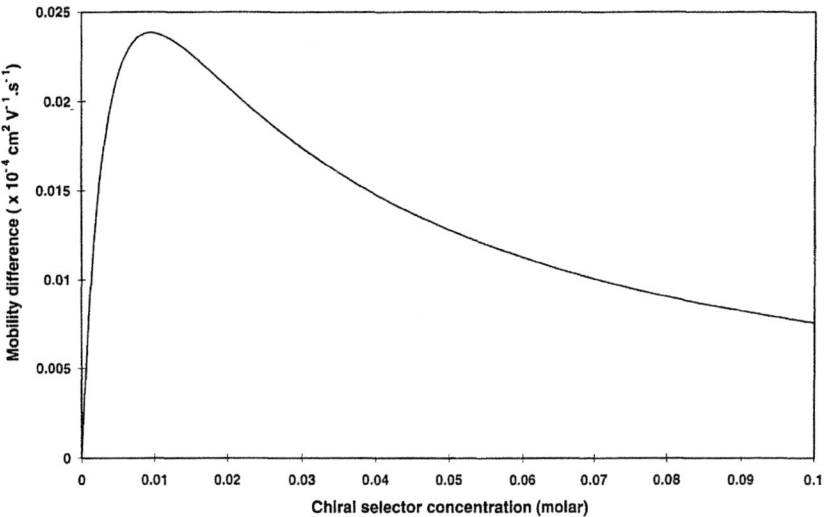

Figure 3.6. Mobility difference between enantiomers A and B using $\mu_0 = 2 \times 10^{-4}$ cm^2 V^{-1} s^{-1} and $\mu_1 = 1 \times 10^{-4}$ cm^2 V^{-1} s^{-1} with $K_1 = 100$ M^{-1} and $K_2 = 110$ M^{-1} using the selector range 0–0.1 M.

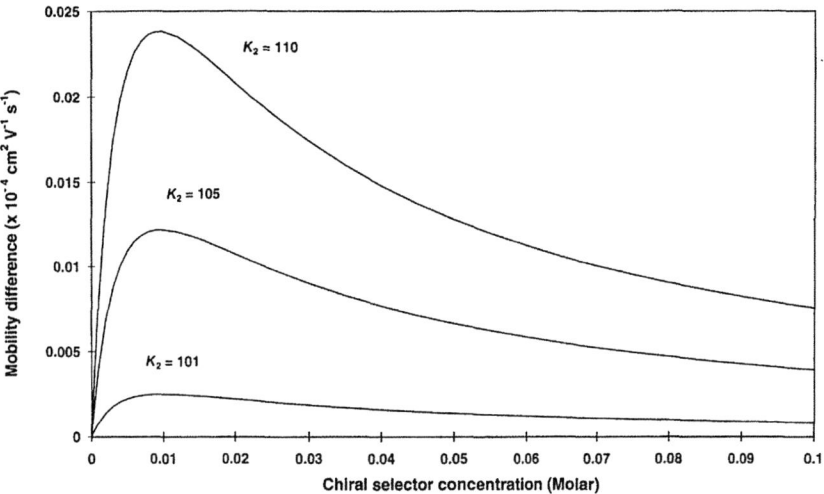

Figure 3.7. Mobility difference between enantiomers A and B using $\mu_0 = 2 \times 10^{-4}$ cm^2 V^{-1} s^{-1} and $\mu_1 = 1 \times 10^{-4}$ cm^2 V^{-1} s^{-1} with $K_1 = 100$ M^{-1} and $K_2 = 101$ M^{-1}, $K_2 = 105$ M^{-1}, and $K_2 = 110$ M^{-1} using the selector range 0–0.1 M.

In each case shown in Figure 3.8 the relative affinities of the two enantiomers are 10% different. The equilibrium constants are: $K_1 = 10$ and $K_2 = 11$; $K_1 = 100$ and $K_2 = 110$; $K_1 = 1000$ and $K_2 = 1100$. These sets of equilibrium constants are used with the parameters $\mu_0 = 2 \times 10^{-4}$ cm^2 V^{-1} s^{-1}; and $\mu_1 = 1 \times 10^{-4}$ cm^2 V^{-1} s^{-1} and the chiral selector concentration range 0–0.1 M. From Figure 3.8 we can see that the apparent mobility difference as a function of the chiral selector concentration is very different for the three sets of equilibrium constants. For the equilibrium constants $K_1 = 1000$ and $K_2 = 1100$ the apparent mobility difference increases very rapidly until a maximum is reached and then drops away quickly. With the equilibrium con-

stants $K_1 = 10$ and $K_2 = 11$ the apparent mobility difference climbs slowly and only levels off at about 100 mM. The maximum apparent mobility difference is the same for each of the three pairs of equilibrium constants. The cases differ in the chiral selector concentration at which the mobility difference is maximised. The chiral selector concentration which maximises the apparent mobility difference is the optimum chiral selector concentration. For the equilibrium constants $K_1 = 1000$ and $K_2 = 1100$ the optimum chiral selector concentration is about 1 mM. For the equilibrium constants $K_1 = 100$ and $K_2 = 110$ the optimum chiral selector concentration is about 10 mM. The greater the absolute affinity of the analytes for the

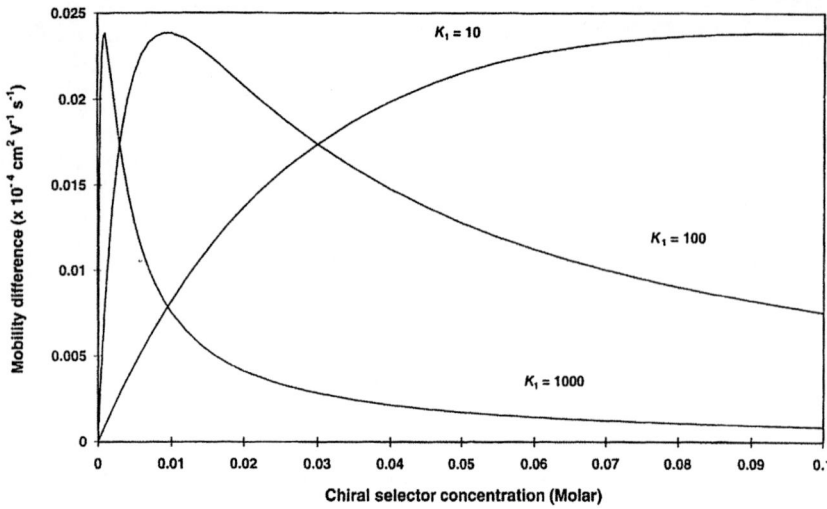

Figure 3.8. Mobility difference between enantiomers A and B using $\mu_0 = 2 \times 10^{-4}$ cm^2 V^{-1} s^{-1} and $\mu_1 = 1 \times 10^{-4}$ cm^2 V^{-1} s^{-1} with the equilibrium constants $K_1 = 10$ M^{-1} and $K_2 = 11$ M^{-1}, $K_1 = 100$ M^{-1} and $K_2 = 110$ M^{-1}, and $K_1 = 1000$ M^{-1} and $K_2 = 1100$ M^{-1} using the range 0–0.1 M.

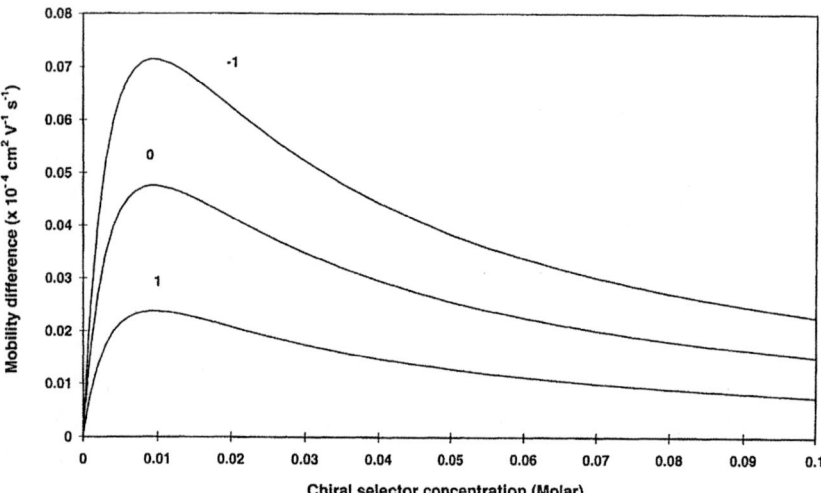

Figure 3.9. Mobility difference between enantiomers A and B with $K_1 = 100$ M^{-1} and $K_2 = 110$ M^{-1} using the chiral selector range 0–0.1 M, with $\mu_0 = 2 \times 10^{-4}$ cm^2 V^{-1} s^{-1} and limiting mobilities of $\mu_1 = 1, 0$, and -1×10^{-4} cm^2 V^{-1} s^{-1}.

chiral selector the smaller the optimum concentration of chiral selector.

It can be seen in equation (3.5) that the apparent mobility difference between the two enantiomers is altered by the difference between the electrophoretic mobilities of the free and bound forms of the analyte. In Figure 3.9 the apparent mobility difference curves generated by three different limiting mobilities are shown.

Each of the curves in Figure 3.9 uses the equilibrium constants $K_1 = 100$ and $K_2 = 110$ along with a value for the electrophoretic mobility of the free analyte of $\mu_0 = 2 \times 10^{-4}$ V^{-1} s^{-1}. The three curves arise from values of the limiting mobility of the complexed analyte, μ_1, of the fol-

lowing: 1, 0, and -1×10^{-4} cm^2 V^{-1} s^{-1}. These different values for the limiting mobilities might arise for example from the following cases : 1) an electrically neutral chiral selector employed with a positively charged analyte; 2) a chiral selector which was fixed in a gel matrix, or a neutral chiral selector of very high mass was used; and 3) a chiral selector carrying charges opposite to that on the analyte was employed. It can be seen in Figure 3.9 that the three curves generated are of the same shape and have the same optimum chiral selector concentration. They differ in the degree of apparent mobility difference generated. In the following chapter it will be seen that cyclodextrins carrying negative

charge can produce some very impressive separations of the enantiomers of positively charged analytes.

3.4.3 An Analytical Solution

Equation (3.5) was investigated by substituting in some realistic parameters and generating graphs showing mobility difference as a function of the chiral selector concentration. Whilst this computer modelling approach is simple and illustrative it is also important to determine a general mathematical solution to the problem.

From the graphs above we are interested in the optimum chiral selector concentration and the apparent mobility difference which is obtainable at the optimum concentration. The answers to these questions related to optimum concentration can be obtained by the use of differential calculus. The optimum chiral selector concentration occurs at the point when the apparent mobility difference has attained its maximum value. At this point the rate of change of electrophoretic mobility with the change in chiral selector concentration is zero. This is stated mathematically in equation (3.6).

$$\frac{d\Delta\mu}{d[\text{C}]} = 0 \qquad (3.6)$$

Following the differentiation of (3.5) it can be shown that equation (3.6) is satisfied when equation (3.7) is true.

$$(K_2 - K_1)(\mu_1 - \mu_2)(1 - K_1 K_2 [\text{C}]^2) = 0 \qquad (3.7)$$

So apart from the trivial solutions $K_2 = K_1$, and $\mu_1 = \mu_0$, the condition is satisfied by equation (3.8).

$$[\text{C}]_{\text{opt}} = \frac{1}{\sqrt{K_1 K_2}} \qquad (3.8)$$

Equation (3.8) shows that the optimum chiral selector concentration is inversely related to the affinity of the analyte for the chiral selector; analytes which have a high affinity for the chiral selector require a low concentration to maximise mobility difference and vice versa. The result shown in equation (3.8) is important as it indicates that we can elevate method development from a trial and error procedure by a consideration of molecular properties. We should not expect a single chiral selector concentration to work for all analytes but should be prepared to examine a wide range of concentrations.

Chromatographia Supplement Vol. 54, 2001

Original

The size of the apparent electrophoretic mobility of the first enantiomer at the optimum chiral selector concentration is found by substituting equation (3.8) into equation (3.3). This gives rise to equation (3.9).

$$\mu_{a(opt)} = \frac{\mu_0 + \mu_1 \sqrt{\frac{K_1}{K_2}}}{1 + \sqrt{\frac{K_1}{K_2}}} \quad (3.9)$$

At the optimum chiral selector concentration the apparent mobility of the first (and also the second) enantiomer depends only upon: the equilibrium constants, the mobility of the free analyte, and the limiting mobility of the analyte-chiral selector complex. Equation (3.9) can be simplified by applying the following approximation. For difficult chiral separations the second equilibrium constant will be of a similar size to the first i.e. $K_2 \approx K_1$. This leads to the simple result shown in equation (3.10).

$$\mu_{a(opt)} \approx \frac{\mu_0 + \mu_1}{2} \quad (3.10)$$

This means that at the optimum chiral selector concentration the apparent mobility of the first enantiomer will be close to midway between the mobilities of the free and complexed forms of the analyte. It is clear that an appreciation of how mobility is changing with changes in chiral selector concentration will be of great benefit in developing chiral separations.

The size of the electrophoretic mobility difference which is obtained at the optimum chiral selector concentration can be obtained by combining equations (3.5) and (3.8). The result can be further simplified by noting that the second equilibrium constant will be some ratio of that of the first i.e. $K_2 = nK_1$. Thus we obtain the result shown in equation (3.11).

$$\Delta\mu_{opt} = \frac{(n-1)(\mu_0 - \mu_1)}{(\sqrt{n} + 1)^2} \quad (3.11)$$

So the maximum mobility difference will depend upon the ratio of the equilibrium constants and difference in the mobilities of the free and complexed forms of the analyte. This is in line with the observation drawn from Figure 3.8 which showed that three pairs of equilibrium constants with the same ratio of K_2/K_1 gave the same maximum mobility differences. Equation (3.11) underlines the importance of selecting operating conditions which increase the ratio of the equilibrium constants and the mobility difference between the free and bound forms of the analyte.

The effect of varying n, the ratio of K_2/K_1, on the maximum mobility difference is shown in Figure 3.10.

From Figure 3.10 we can see that when n is close to one the relationship between n and the maximum mobility difference is approximately linear and is proportional to the term $(n-1)$. As the value of n becomes larger the term $(n^{0.5} + 1)^2$ becomes more important and the rate of increase in the maximum mobility difference drops off.

3.4.4 Interpretation of Previous Data

The modelling and analysis given in the preceding section can give some answers to some of the apparent contradictions and strange patterns seen in the experimental data. Whether increasing the chiral selector concentration leads to an increase or decrease in resolution will depend upon the concentration in relation to the affinity. The important factor is whether the change in the chiral selector concentration takes the concentration towards or beyond the optimum value of $1/(K_1/K_2)^{0.5}$. From Figure 3.8 we can see that an increase in chiral selector concentration from 1 to 2 mM would lead to an improvement in resolution with the equilibrium constants $K_1 = 100$ and $K_2 = 110$. The same selector concentration change from 1 to 2 mM would, however, lead to a decrease in resolution with the equilibrium constants $K_1 = 1000$ and $K_2 = 1100$.

Figure 3.8 can also help us to understand the influence of changing the concentration of organic solvent in the buffer. It is often assumed that the complexes which are produced by cyclodextrins arise from the inclusion of the hydrophobic portion of the analyte into the hydrophobic cyclodextrin cavity. If an organic solvent such as acetonitrile is added to the buffer this will make the environment around the cyclodextrin less polar. This reduction in bulk buffer polarity will tend to reduce the attraction of a hydrophobic analyte for the cyclodextrin cavity i.e. K_1 and K_2 will be reduced. Whether the reduction in K_1 and K_2 caused by the organic solvent leads to an increase or decrease in resolution will again depend on the original concentration and the size of the affinity. In Figure 3.8 we can see that with a chiral selector concentration of 10 mM reducing the equilibrium constants from $K_1 = 1000$ and $K_2 = 1100$ to $K_1 = 100$ and $K_2 = 110$ would lead to an increase in mobility difference and so resolution. Conversely reducing the equilibrium constants from $K_1 = 100$ and $K_2 = 110$ to $K_1 = 10$ and $K_2 = 11$ would lead to a decrease in mobility difference and resolution. The direction of change in resolution depends upon whether the chiral selector concentration in the absence of organic solvent is above or below the optimum value.

3.4.5 Experimental Support

In the preceding discussion it was shown that trends in the literature data could be explained by the equilibrium model. The

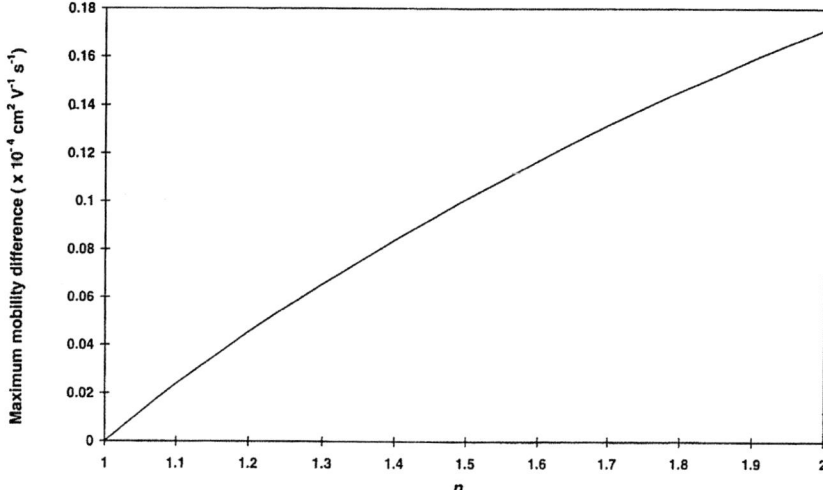

Figure 3.10. Mobility difference at the optimum chiral selector concentration as a function of n, the ratio of the equilibrium constants.

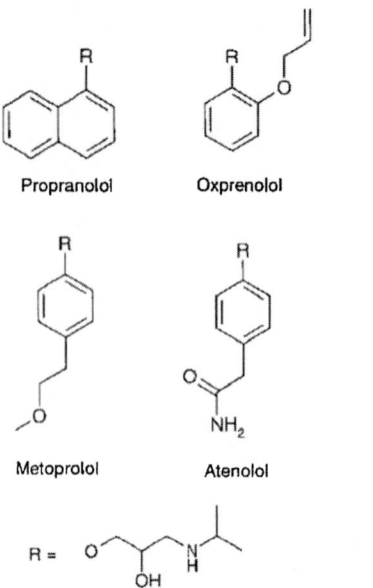

Figure 3.11. The β-blockers propranolol, oxprenolol, metoprolol, and atenolol.

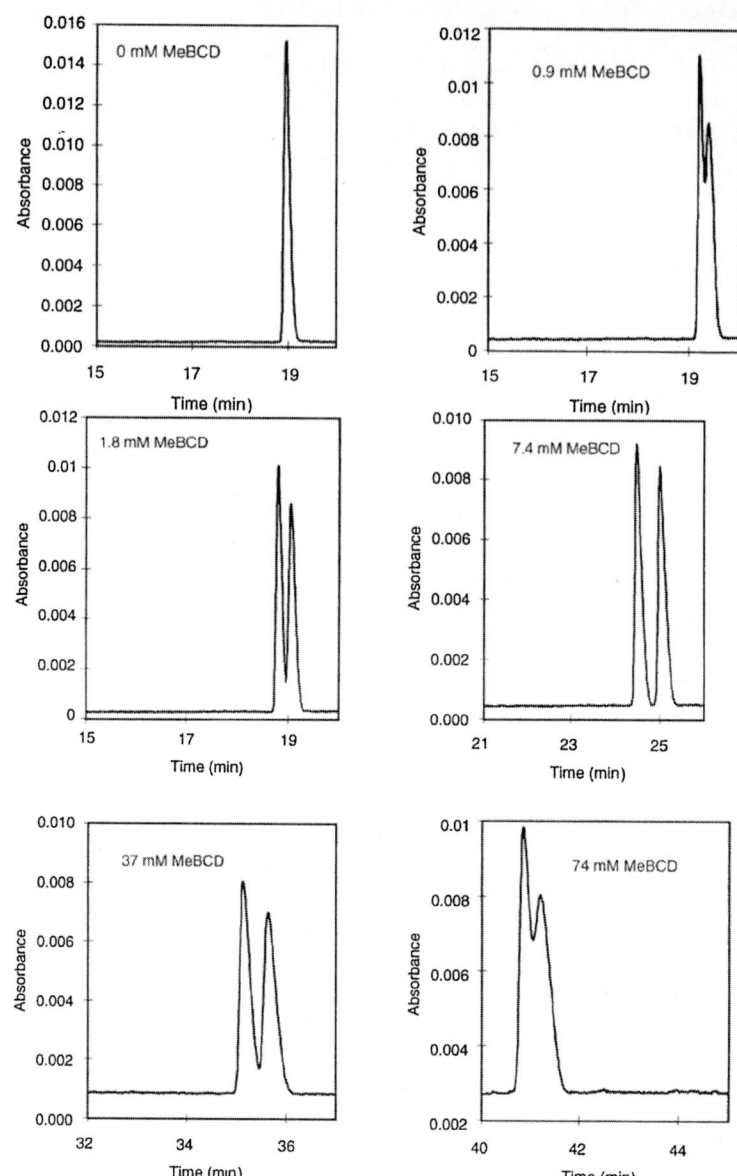

Figure 3.12. The resolution of propranolol enantiomers at six different DMBCD concentrations.

next stage in the development was to determine whether the model was limited to an explanation of existing data or whether it could correctly predict the results of further experimentation. Of particular interest was whether the hydrophobicity of the analyte could be related to equilibrium constants and hence optimum selector concentration. To test the model a family of related compounds were selected. The β-blockers were chosen as they are an important class of pharmaceutical compound and have good solubility and UV detection properties. The work of Fanali [2] had shown that good results could be obtained by using the β-blocker propranolol (Figure 3.11) along with native β-cyclodextrin and methyl substituted β-cyclodextrins.

The cyclodextrins chosen to test the model were the parent β-cyclodextrin (BCD) and dimethyl-β-cyclodextrin (DMBCD) [11]. The latter β-cyclodextrin had the 2-, 3-, and 6- hydroxy groups of the sugar units partially substituted with methoxy ones. The average degree of substitution was 1.8 methoxy units per glucose ring. The experiments were carried out in a low pH lithium phosphate buffer to reduce the electroosmotic flow. The first results were obtained using propranolol with DMBCD [11]. The influence of the DMBCD can be seen in Figure 3.12 which shows the separations between the propranolol enantiomers obtained at six different concentrations.

Initially the separation increases with increasing DMBCD concentration but after reaching a maximum then declines. By carrying out spiking experiments (R)-(+) propranolol was shown to be the second migrating peak and hence the enantiomer with the highest affinity for DMBCD. The results shown in Figure 3.12 are in line with those expected from the model. The measured electrophoretic mobility differences between the propranolol enantiomers are shown in Figure 3.13. The shape of the line shown in Figure 3.13 is of a similar form to the computer generated plots given before and therefore supports the equilibrium model.

The greatest electrophoretic mobility differences were recorded with concentrations of 3.7 mM and 7.4 mM. The shape of the curve in Figure 3.13 implies that a higher electrophoretic mobility difference would be obtained with an intermediate DMBCD concentration of about 5.5 mM. The maximum resolution obtained is greater than baseline and so could be expected to be sufficient for the reliable quantification of a minor enantiomer. The high resolution between the propranolol enantiomers is partly obtained because of the sharpness of the peaks. Typical measured efficiencies of the peaks due to the individual enantiomers are 115,000 theoretical plates.

The experiment was repeated using propranolol with β-cyclodextrin (BCD) as the chiral selector. In this case 4 M urea had to be added to the buffer in order to solubilise the cyclodextrin. Figure 3.14 shows the measured electrophoretic mobility differences of the propranolol enantiomers as a function of BCD concentration.

Comparing the curve shown in Figure 3.14 with that from Figure 3.13 shows two interesting features: a) the optimum chiral selector concentration, and b) the maximum mobility difference. The optimum BCD concentration is about 14 mM as opposed to a value of about 5.5 mM for DMBCD. This concentration difference shows that the affinity of the propranolol enantiomers for BCD is less than half that for DMBCD. The maximum mobility difference obtained with BCD is only about 0.01×10^{-4} cm^2 V^{-1} s^{-1} i.e. about half of the value obtained with DMBCD. The two maximum mobility difference values show that differential affinity between the two enantiomers is much greater with DMBCD than BCD. The reasons for the marked differences in performance between BCD and DMBCD are not known but are presumably related to the chemical or structural changes at the rim of the cyclodextrin cavity. The resolution obtained between the enantiomers of propranolol with the optimum concentration of BCD is shown in Figure 3.15.

The importance of analyte hydrophobicity was investigated by repeating the propranolol experiments with DMBCD using the β-blockers oxprenolol, metoprolol, and atenolol [13] (see Figure 3.11). The log P values (octanol-water partition coefficient) are shown in Table 3.I [14].

From Table 3.I it can be seen that propranolol is the most hydrophobic β-blocker, atenolol the least and that metoprolol and oxprenolol have intermediate values. One of the assumptions about the behaviour of cyclodextrins as chiral selectors is that the hydrophobic portion of the analyte molecule is included into the hydrophobic cyclodextrin cavity. As a

Table 3.I. Hydrophobicity of β-blockers

β-blocker	log P
Atenolol	0.23
Metoprolol	2.15
Oxprenolol	2.18
Propranolol	3.65

▷

Figure 3.15. The resolution of the propranolol enantiomers at the optimum concentration of BCD.

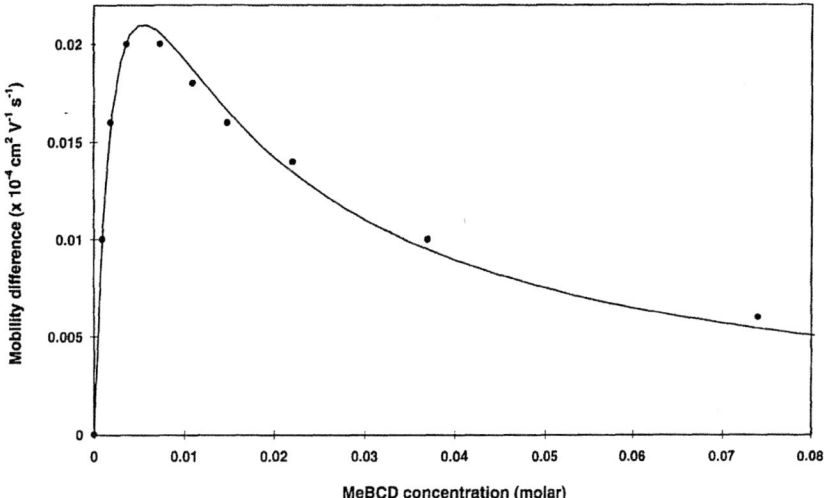

Figure 3.13. Measured apparent mobility difference between propranolol enantiomers as a function of the concentration of DMBCD.

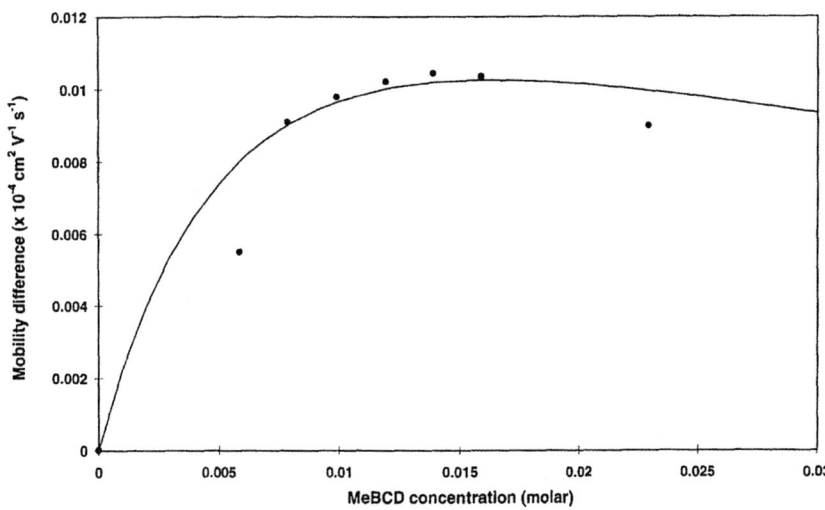

Figure 3.14. Measured apparent mobility difference between propranolol enantiomers as a function of the concentration of BCD.

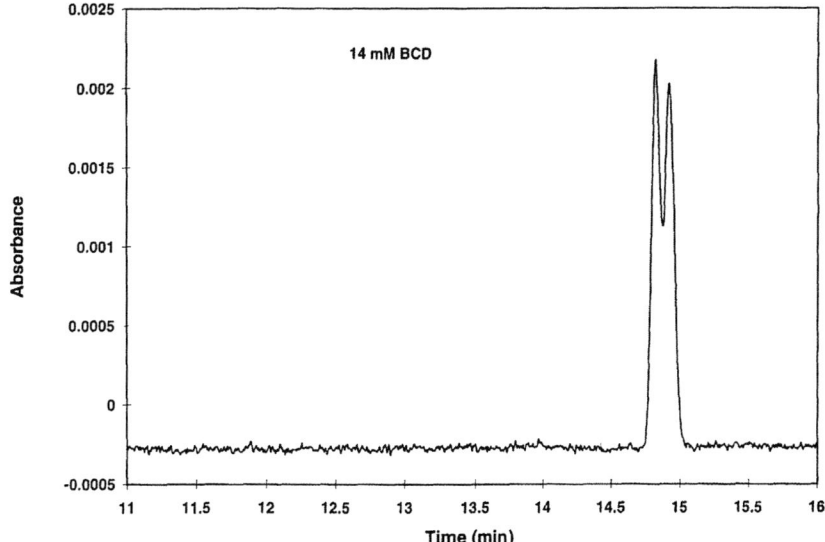

first approximation we can make the assumption that the affinity of the analyte for the cyclodextrin will be dominated by hydrophobic interaction. On this basis we would expect atenolol, which is the least hydrophobic analyte, to have the least affinity for DMBCD. Conversely propranolol the most hydrophobic analyte should have the greatest affinity for DMBCD.

As the optimum chiral selector concentration is inversely related to the affinity, propranolol should have the lowest optimum chiral selector concentration and atenolol the highest. oxprenolol and metoprolol could be expected to have optimum chiral selector concentrations intermediate between those of propranolol and atenolol.

Figure 3.16 shows the separation of the enantiomers of atenolol that were obtained using DMBCD concentrations of 0, 7, 15, 37, and 74 mM.

The trend is the same as that observed with propranolol – the resolution initially increases with increasing DMBCD concentration but then reaches a maximum value. At higher concentrations of DMBCD the resolution of the atenolol enantiomers declines slightly. The major difference from propranolol is that the optimum DMBCD concentration is much higher with atenolol, as expected from the difference in hydrophobicity. The concentration of atenolol used was much lower than that of propranolol and this is the reason for the symmetrical peaks. The system efficiency is independent of the DMBCD concentration with values of 175,000 plates being recorded with concentrations of both 0 and 37 mM. The high and unchanged efficiencies indicate that the exchange between free and complexed forms of atenolol is rapid on the timescale of the CE experiment.

Figure 3.17 shows the separation of oxprenolol enantiomers obtained by using DMBCD concentrations of 0, 11, 22, 37, and 74 mM.

The pattern seen with propranolol and atenolol is again observed with oxprenolol with an optimum DMBCD concentration being seen. It is again clear that the optimum DMBCD concentration is considerably higher than that recorded with propranolol. The sharpness of the peaks is again unaltered by the change in chiral selector concentration. The maximum resolution of oxprenolol enantiomers is slightly less than that achieved with atenolol and significantly less than that achieved with propranolol.

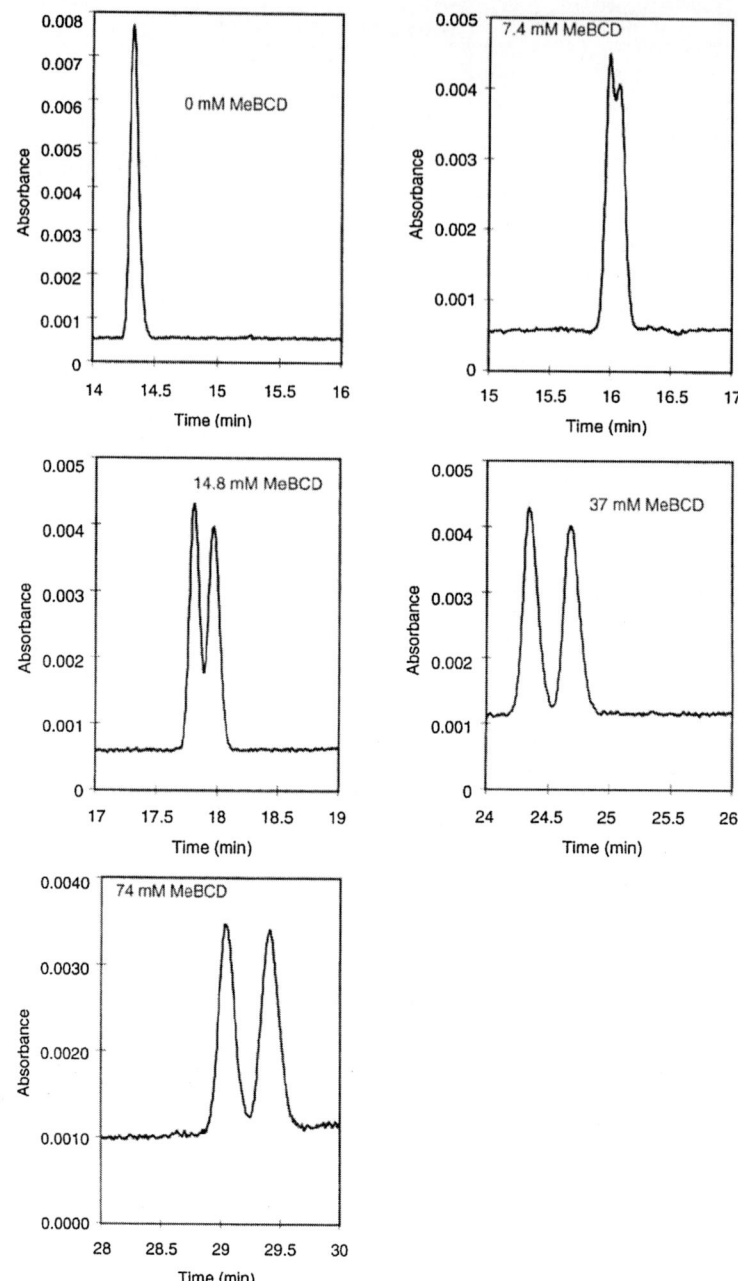

Figure 3.16. The resolution of the atenolol enantiomers at five different DMBCD concentrations.

The measured mobility differences as a function of the DMBCD concentration is shown for all four β-blockers in Figure 3.18.

From Figure 3.18 it can be seen that, as expected, the optimum chiral selector concentration varies with the hydrophobicity of the analyte. Propranolol the most hydrophobic β-blocker requires the lowest concentration of DMBCD and atenolol the least hydrophobic requires the highest. Metoprolol which has intermediate hydrophobicity has an intermediate optimum chiral selector concentration. The

only inconsistency in this pattern is provided by oxprenolol which has similar hydrophobicity to metoprolol but a similar optimum concentration to atenolol. A possible explanation for this is the importance of steric factors. Oxprenolol has a 2 rather than a 4 substitution pattern and this will lead to an inclusion complex with a higher degree of steric crowding. The steric crowding could be expected to lead to a lower affinity than might be expected from hydrophobicity alone. So whilst hydrophobicity is a useful way of ranking the likely affinities for a family of compounds it only gives a first order approximation. The other feature of interest is the large degree of variation in the apparent mobility difference obtained at the optimum chiral selector concentration. The maximum mobility difference does not correlate with the hydrophobicity of the analyte and an examination of the 2 dimensional structure of the analyte gives few clues either.

In the preceding section it was noted that the addition of organic solvent could either increase or decrease resolution. The influence of the organic solvent was rationalised in terms of the change in the equilibrium constants, and whether the chiral selector concentration was above or below the optimum. Fanali [2] had showed that with 40 mM of BCD the addition of methanol improved the resolution of propranolol enantiomers. This can be explained if the concentration of 40 mM is greater than the optimum in the absence of organic solvent. This is consistent with the data shown in Figure 3.14. An interesting test of the model would therefore be to check that with the chiral selector concentration below the optimum, the addition of organic solvent would lead to a decrease in resolution.

The idea was checked by adding different amounts of methanol to buffer systems containing 3.7 mM of DMBCD. This concentration had been shown to below the optimum for an entirely aqueous buffer system [11]. The addition of methanol could be expected to reduce the affinity of propranolol for DMBCD and so make the concentration of 3.7 mM even less optimal. The resolution of propranolol enantiomers obtained using buffers with constant ionic strength and cyclodextrin concentration but with 0, 5 and 19% (v/v) of methanol is shown in Figure 3.19 [15].

The decrease in resolution with increasing concentration of methanol can be seen clearly in Figure 3.19. It is also clear that there is an increase in migration time with methanol concentration. The increase in the migration time is thought to be due to the increase in viscosity of mixed methanol/water systems. The change in apparent mobility difference as a function of methanol concentration is shown in Figure 3.20.

From Figure 3.20 it can be seen that increasing the methanol concentration from 0 to 19 % leads to a reduction in the apparent mobility difference by nearly 50%. A similar trend was also observed by the use

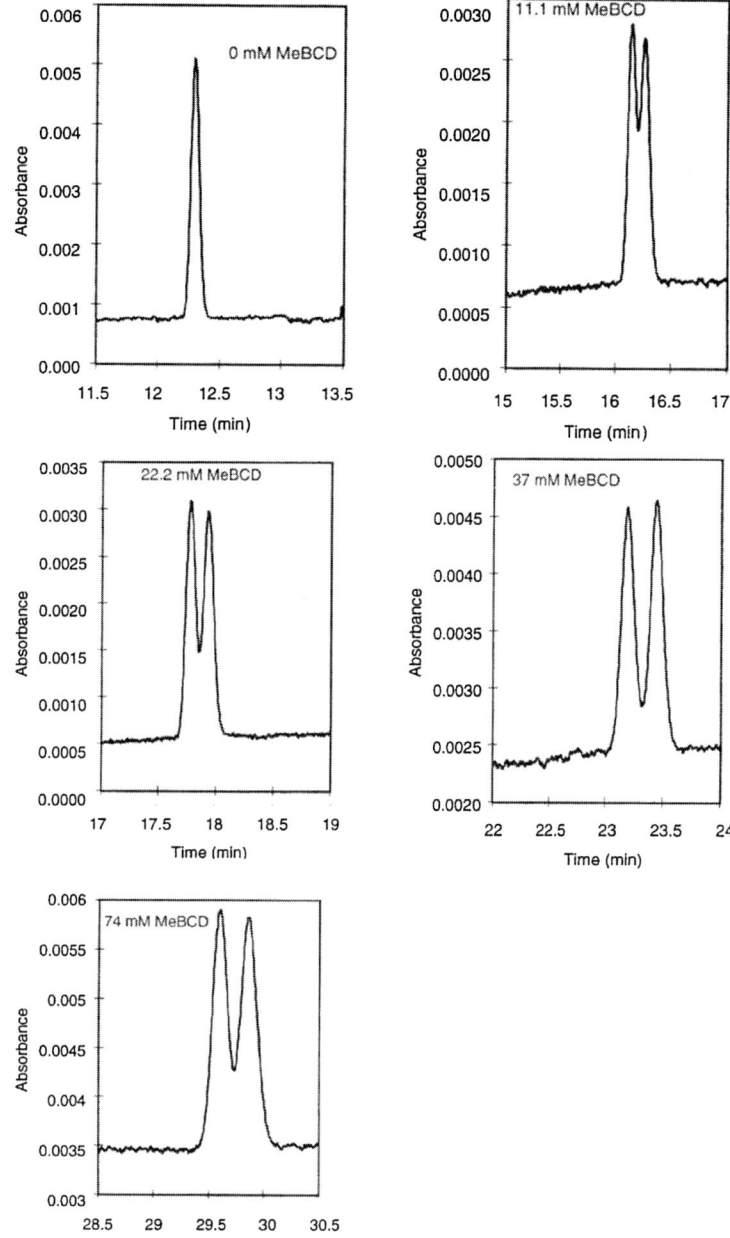

Figure 3.17. The resolution of the oxprenolol enantiomers at five different DMBCD concentrations.

of acetonitrile rather than methanol [15]. Figure 3.21 shows the change in the separation of the enantiomers of propranolol as the proportion of acetonitrile increases from 0 to 15%.

Two important differences between the use of acetonitrile and methanol as solvents were noted. Volume for volume acetonitrile causes a much greater loss in resolution, with about 8% giving the same drop in apparent mobility difference as 19% methanol. In addition the migration times decreased with acetonitrile (a combination of increased apparent mobility and reduced viscosity). The change in apparent mobility difference as a function of acetonitrile concentration is shown in Figure 3.22.

Acetonitrile appears to cause a much greater reduction in the equilibrium constants than methanol with the addition of 15% leading to a total loss of resolution.

The influence of organic solvent and the optimum chiral selector concentration has been examined more fully by other workers. Ferguson, Goodall, and Loran [16] investigated the binding between BCD and enantiomers of the fungicide

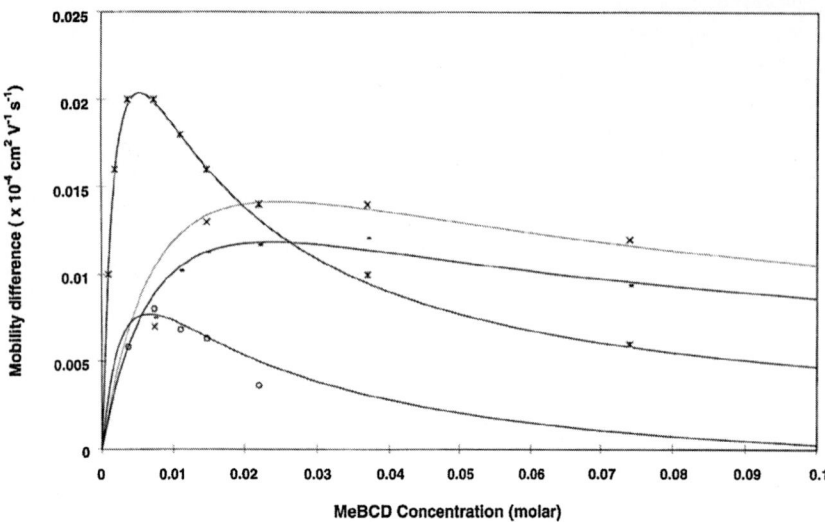

Figure 3.18. Measured mobility difference as a function of DMBCD concentration for the β-blockers propranolol (*), atenolol (x), oxprenolol (–), and metoprolol (o).

Table 3.II. Influence of acetonitrile on the optimum BCD concentration [3.16]

% Acetonitrile	0	5	10	15
$[CD]_{opt}$	0.8	1.6	2.3	6.2

the addition of 0.1% cyclohexanol to the buffer. The change in the observed binding between BCD and tioconazole is consistent with the known affinity of cyclohexanol for BCD.

3.4.6 Resolution

In the previous section we saw how the simple equilibrium model could be used to describe the apparent electrophoretic mobility difference between the pair of enantiomers. It was also shown that the apparent mobility difference is maximised at a chiral selector concentration which is inversely related to the affinities of the analyte for the chiral selector. In the consideration of equation (3.1) it was shown that there are terms other than the electrophoretic mobility difference which control resolution. In particular we can see that the resolution is related to $\Delta\mu_{ep}$ (the mobility difference), μ_{ep} (the electrophoretic mobility), and μ_{eo} (the electroosmotic ability) as shown in equation (3.12).

$$R_S \propto \frac{\Delta\mu_{ep}}{(\mu_{ep} + \mu_{eo})^{0.5}} \qquad (3.12)$$

So it is clear that we not only need to maximise the term $\Delta\mu_{ep}$, but also to minimise the term $(\mu_{ep} + \mu_{eo})^{0.5}$. Under normal conditions the electroosmotic flow is towards the cathode and so for cationic analytes $(\mu_{ep} + \mu_{eo})$ will always be positive. For cations it is therefore sensible to try and minimise the electroosmotic mobility so as to maximise resolution. For anionic analytes it is possible to obtain good resolution when the electrophoretic mobility is similar in magnitude but opposite in direction to the electroosmotic mobility. In theory infinite resolution is obtainable when $\mu_{ep} = -\mu_{eo}$. This mobility matching is however of limited practical utility as the analysis time would also be infinite! Strategies for controlling and reducing electroosmotic mobility as a means of improving resolution are considered in more detail in Chapter 4.

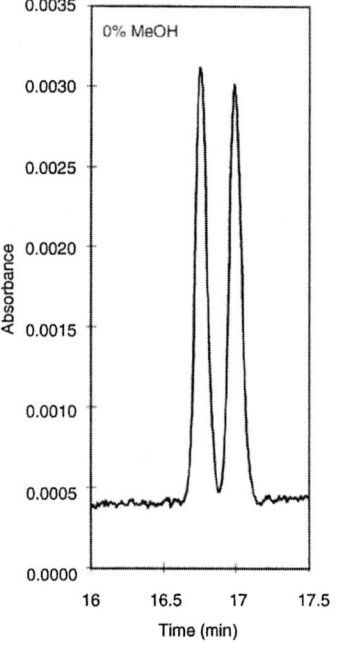

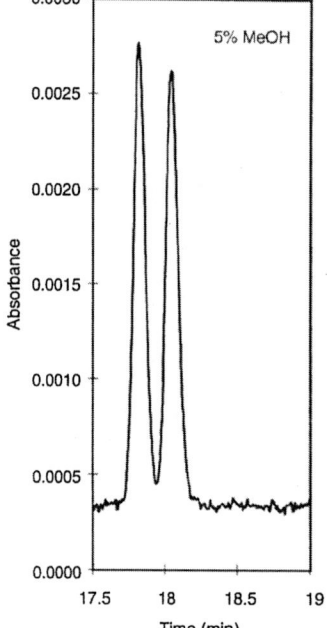

Figure 3.19. Legend see next page.

tioconazole using different concentrations of acetonitrile. They found the optimum chiral selector concentration to increase with increasing concentration of acetonitrile as shown in Table 3.II. The addition of only 5% of acetonitrile leads to a doubling of the concentration of BCD required to optimise the separation. The measurement of the equilibrium constants showed a negative linear relationship between log K and the percentage of acetonitrile. It was also noted that the rate of change in log K was about twice as great with acetonitrile as methanol.

The data on tioconazole also show that both acetonitrile and methanol do not affect the selectivity of the separation, with the ratios of the equilibrium constants showing little change with changing concentration. The use of organic solvent is considered in greater detail in Chapter 4.

The basic concept of the cyclodextrins acting via the formation of transient inclusion complexes has been supported by the addition of competing species. Goodall et al. [17] showed that the equilibrium constants for tioconazole enantiomers and BCD were reduced by a factor of six by

3.5 Extended Theoretical Models

The discussion up until now has considered the case of single forms of the two enantiomers interacting with a single chiral selector to give two complexes with the same limiting mobilities. This is clearly a good first order approximation as it provides a good fit to the data in many cases, as has been well illustrated in the preceding sections. It is obvious however that the approximation of single forms and the same limiting mobilities is a simplification which cannot be expected to hold, or to be useful, in all cases. It is therefore appropriate that several of the other interesting possibilities are also considered.

The first possibility is that the two enantiomers can interact with the chiral selector to form complexes which have different limiting mobilities. The second possibility is that of multiple equilibria, with the analyte enantiomers forming a complexes with more than one chiral selector. And finally the analyte enantiomers can exist in more than one form and these forms can interact to different extents with the chiral selector. An example of this third possibility is protonation equilibria involving the analyte. In this situation the analyte enantiomers are partially charged and the protonated and de protonated forms can interact to different extents with the chiral selector.

3.5.1 Different Limiting Mobilities

In the discussion on physical and mathematical models earlier in this chapter we considered the case of the two enantiomer-chiral selector complexes having the same limiting mobility. Equal limiting mobilities for the two complexes is of course a simplifying assumption and we shall now consider the case where they differ [18]. The difference in limiting mobilities of the two enantiomer-chiral selector complexes can be expressed mathematically as $\mu_1 \neq \mu_2$ (where μ_1 is the limiting mobility of the complex formed by the first enantiomer and μ_2 the limiting mobility of the complex formed by the second enantiomer). In effect the inclusion complexes formed by the two enantiomers are different diastereoisomeric forms and so could have different properties. The mobility difference between the two enantiomer-selector complexes can arise as whilst electrophoretic mobility is dominated by

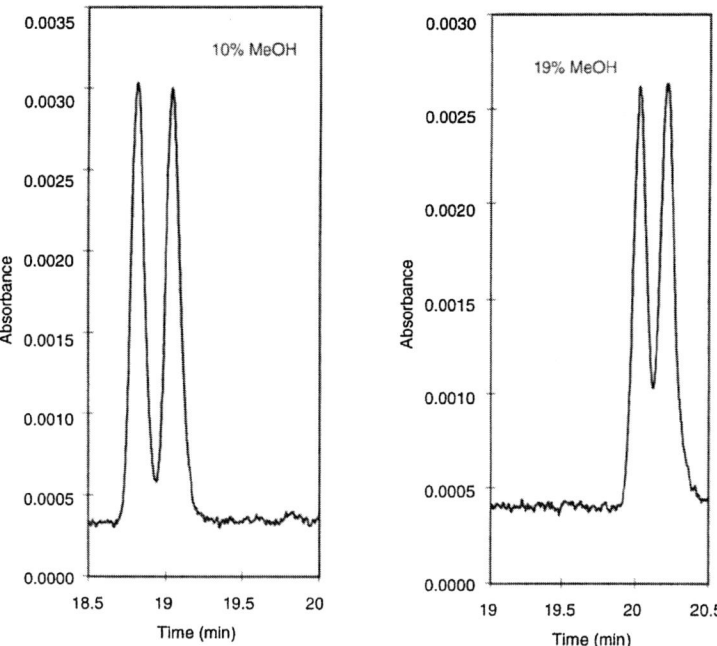

Figure 3.19. The resolution of the propranolol enantiomers at four different methanol concentrations.

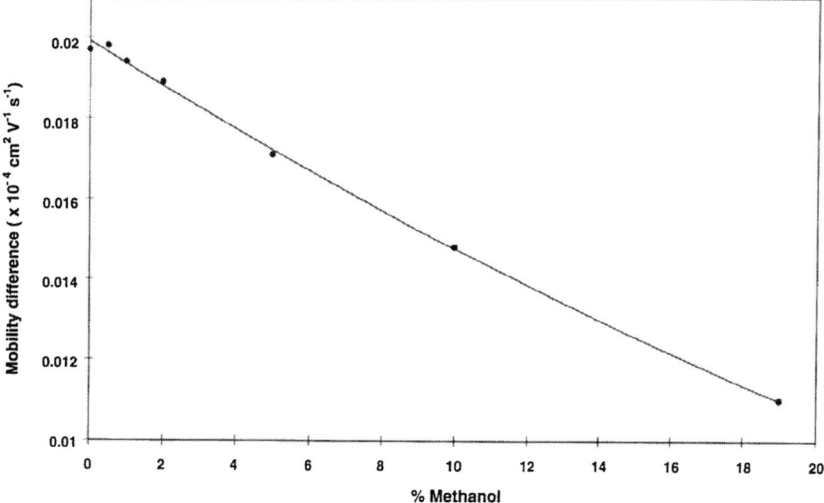

Figure 3.20. The change in the apparent mobility difference of the propranolol enantiomers as a function of methanol concentration.

the size and charge of the analyte, the shape of the analyte is also important. The shape effect is small but as was noted in Chapter 2 can be sufficient to induce separation of isomeric species. The shape difference could arise for example by the two enantiomers adopting a slightly different orientation in the cavity or differing in their depth of inclusion.

The significance of a difference in limiting mobility is again easily explored by the use of mathematical modelling. The graphs in Figure 3.23 and 3.24 are constructed using different pairings of the fol-

lowing equilibrium and limiting mobility parameters: $K_1 = 100$ M^{-1} and $K_2 = 130$ M^{-1}, $\mu_0 = 2 \times 10^{-4}$ cm^2 V^{-1} s^{-1}, the two limiting mobilities have the values of 1.05 and 1.0×10^{-4} cm^2 V^{-1} s^{-1}.

In Figure 3.23 the enantiomer with the lower affinity for the chiral selector also gives the complex with the greater limiting mobility. In Figure 3.23 the apparent mobilities of the two complexes do not converge at very high chiral selector concentration but instead the difference in apparent mobilities tends towards that of the differences in the limiting mobilities. Such

Chromatographia Supplement Vol. 54, 2001

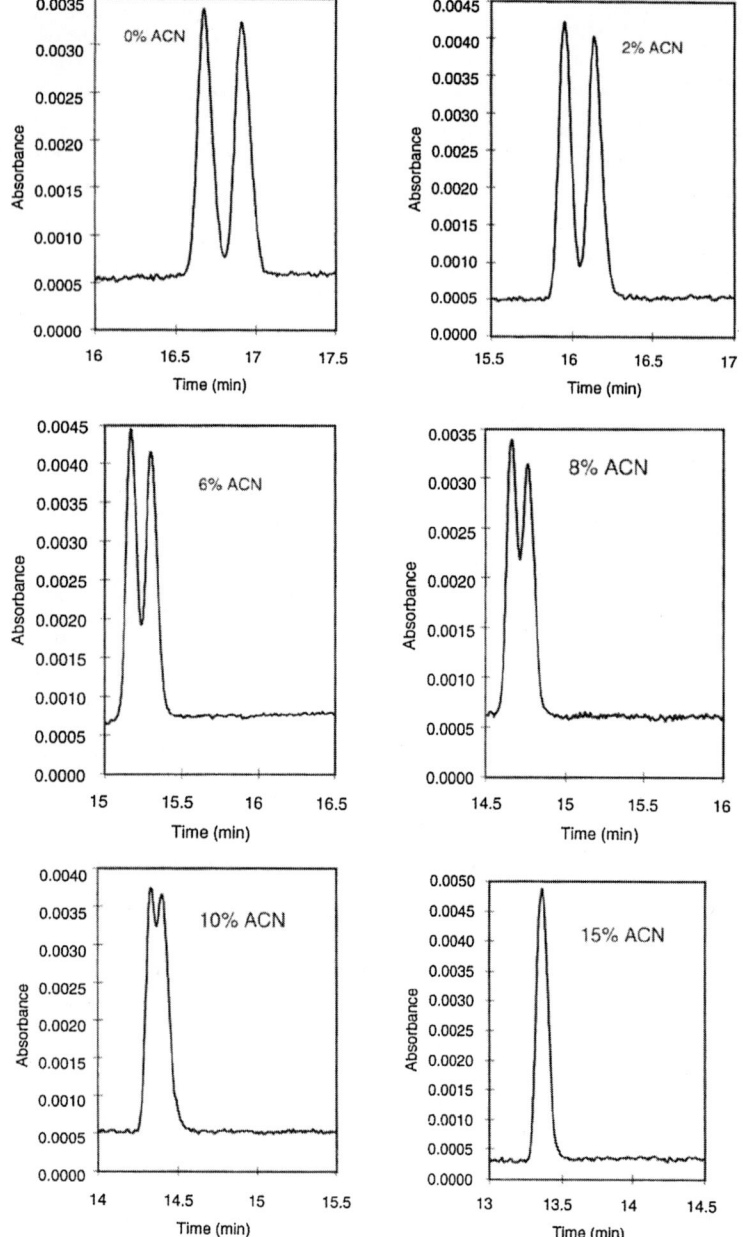

Figure 3.21. The resolution of the propranolol enantiomers at different acetonitrile concentrations.

dominated by affinity differences, whereas at high concentration limiting mobility differences are important.

In Figure 3.25 the modulus of the mobility difference is plotted against the chiral selector concentration for the sets of conditions given in Figures 3.23 and 3.24.

It can be seen from Figure 3.25 that the change in migration order leads to a minimum value for mobility difference in addition to the maxima. In the previous section it was mentioned that at infinite chiral selector concentration the apparent mobility difference will simply be the difference in the limiting mobilities of the two enantiomers. From this it follows that the two curves seen in Figure 3.25 will coincide at infinite chiral selector concentration. The modulus is used as resolution is a scalar rather than a vector quantity. A change in migration order is a potentially useful event. For a material which is enantiomerically very pure we may wish to change the positioning of the minor enantiomer such that it is not obscured by the major enantiomer or other related substances.

The situation can also be analysed mathematically [20] with the electrophoretic mobility difference described by equation (3.13).

see below Eq. (3.13)

Equation (3.13) is clearly more complex than the equivalent equation for equal limiting mobilities. The complexities are compounded when we derive the equation which describes the optimum chiral selector concentration. The optimum concentration is again obtained by differentiation of equation (3.13) and finding the conditions which satisfy $d(\Delta\mu_{ep})/d[C] = 0$. The maxima and minima are given in equation (3.14).

see below Eq. (3.14)

3.5.2 Equilibria Involving more than one Chiral Selector

The model considered up to this point has covered the equilibrium between analyte and chiral selector to form a 1:1 complex. This situation will not always exist and there are two alternatives: the analyte complexing with more than one chiral selector molecule of the same type; and the analyte complexing with more than one type of chiral selector.

The interaction of the analyte with two or more cyclodextrin molecules was pro-

differences in limiting mobilities are relatively uncommon but have been observed with the enantiomers of dansyl phenyalanine using hydroxypropyl-β-cyclodextrin at pH 2.75 [19].

In Figure 3.24 the enantiomer with the lower affinity for the chiral selector gives the complex with the lower limiting mobility. In Figure 3.24 we see a more interesting result with the enantiomers changing migration order with changing chiral selector concentration. The change in migration order occurs because at low concentration the apparent mobilities are

$$\Delta\mu_{ep} = \frac{[C]\{K_1(\mu_1 - \mu_0) - K_2(\mu_2 - \mu_0) + K_1K_2[C](\mu_1 - \mu_2)\}}{1 + [C](K_1 + K_2) + K_1K_2[C]^2} \quad (3.13)$$

$$[C]_{opt} = \frac{(\mu_1 - \mu_2) \pm (K_1 - K_2)\sqrt{\{(\mu_2 - \mu_0)(\mu_1 - \mu_0)/K_1K_2\}}}{K_1(\mu_2 - \mu_0) - K_2(\mu_1 - \mu_0)} \quad (3.14)$$

posed by Sänger-van de Griend et al. [21] to explain results obtained with some tetrapeptide enantiomers. The tetrapeptide, Tyr-Arg-Phe-Phe-NH₂, has four asymmetric centres and so there are eight possible enantiomeric pairs. The pairs of enantiomers were resolved by the use of DM-β-CD. For two of the enantiomeric pairs (LLLL-DDDD, and LDDL-DLLD) the mobility difference varied with concentration in the normal way with a clear optimum chiral selector concentration. For the other six pairs of enantiomers the situation was more complex and the data could not be fitted to the simple model described earlier in this chapter.

The results for three of the pairs (LDLL-DLDD, LDDL-DLLD, and LLDL-DDLD) were particularly interesting as the mobility difference did not decrease in the normal way after reaching the maximum value. In these cases the mobility difference rapidly reached a maximum and then stayed on a plateau at this level, or started to increase again at a much lower rate. For one of the enantiomer pairs at least there was a shallow minimum in the graph of mobility difference vs concentration.

As the tetrapeptide contains three aromatic amino acids the authors considered the possibility of complexes containing two or more cyclodextrin molecules in addition to the 1:1 species. By mathematical modelling they demonstrated that theoretical curves which matched the experimental ones could be obtained. Theoretical mobility difference curves which gave plateau values or shallow minima could be obtained by considering 2:1 and 3:1 complexes.

3.5.3 Equilibria Involving other Species

The models described above considered single chemical forms of the enantiomers interacting with the chiral selector to form equilibrium complexes. The analyte enantiomers may however also be involved in other equilibrium reactions other than with the chiral selector. The most obvious case of additional interactions is seen where the enantiomers are also involved in acid-base equilibria. The analyte enantiomers can exist in either protonated or de-protonated forms and these can have different affinities for the chiral selector. This situation has been analysed and studied in great detail by Rawjee, Staerk, and Vigh who considered acidic analytes [22]. The workers considered how the chiral se-

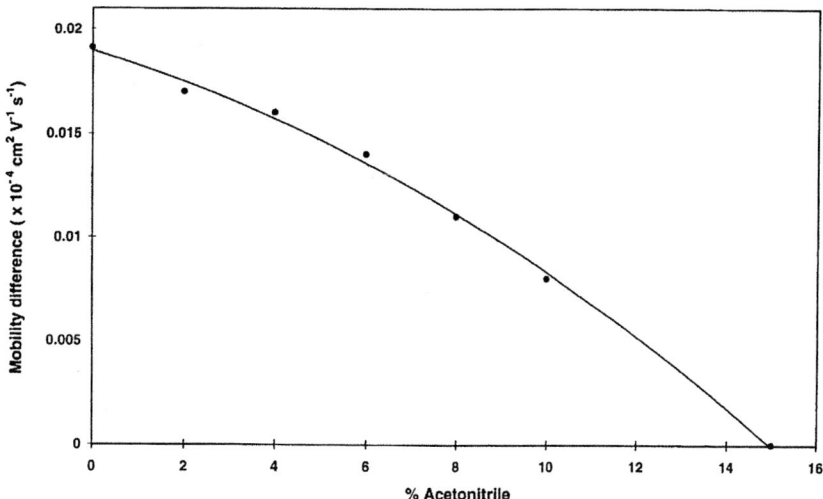

Figure 3.22. The change in the apparent mobility difference of the propranolol enantiomers as a function of acetonitrile concentration.

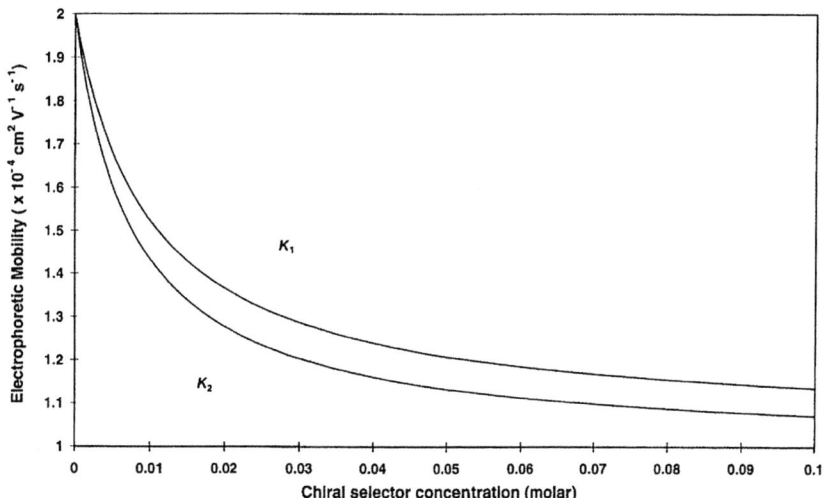

Figure 3.23. The apparent mobilities of enantiomers A and B using $\mu_0 = 2 \times 10^{-4}$ cm² V⁻¹ s⁻¹, $\mu_1 = 1.05 \times 10^{-4}$ cm² V⁻¹ s⁻¹, and $\mu_2 = 1 \times 10^{-4}$ cm² V⁻¹ s⁻¹, with $K_1 = 100$ M⁻¹ and $K_2 = 130$ M⁻¹.

paration selectivity (the ratio of the apparent mobilities of the two enantiomers) could vary as a function of both the pH and the cyclodextrin concentration.

For acids the chiral separation selectivity, $A_{r/s}$, is given in equation (3.15) [22].

see below Eq.   (3.15)

Where μ_0 is the mobility of the free analyte anion, μ_{RCD^-} is the mobility of the complexed R enantiomer anion, and

K_{RCD^-} is the equilibrium constant for the formation of the R enantiomer anion-cyclodextrin complex. K_{HRCD} is the equilibrium constant for the neutral R enantiomer - cyclodextrin complex, K_H is the acid dissociation complex of the free analyte, and [CD] and [H₃O⁺] the concentrations of cyclodextrin and hydroxonium ion respectively. For basic analytes an analogous form of equation (3.15) may also be derived [23].

$$A_{R/S} = \frac{1 + \dfrac{\mu_{RCD^-}}{\mu_0} \cdot K_{RCD^-} \cdot [CD]}{1 + \dfrac{\mu_{SCD^-}}{\mu_0} \cdot K_{SCD^-} \cdot [CD]} \cdot \frac{1 + K_{SCD^-}[CD] + \dfrac{[H_3O^+]}{K_H} \cdot (1 + K_{HSCD}[CD])}{1 + K_{RCD^-}[CD] + \dfrac{[H_3O^+]}{K_H} \cdot (1 + K_{HRCD}[CD])}$$

(3.15)

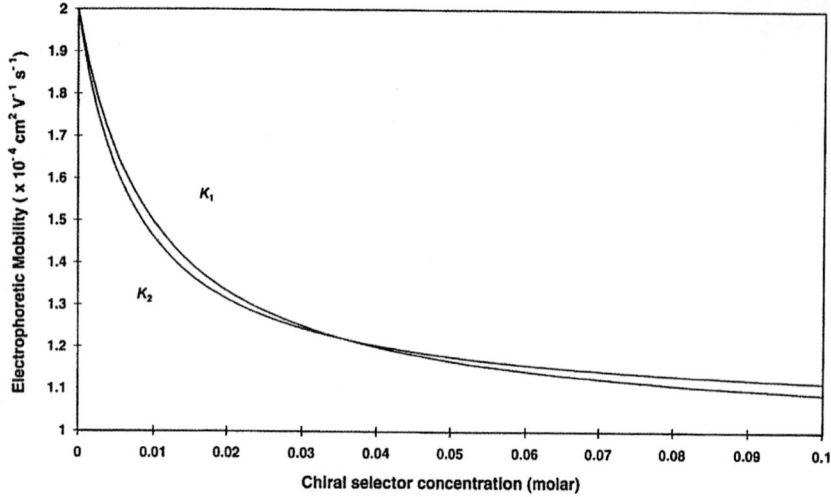

Figure 3.24. The apparent mobilities of enantiomers A and B using $\mu_0 = 2 \times 10^{-4}$ cm^2 V^{-1} s^{-1}, $\mu_1 = 1 \times 10^{-4}$ cm^2 V^{-1} s^{-1}, and $\mu_2 = 1.05 \times 10^{-4}$ cm^2 V^{-1} s^{-1}, with $K_1 = 100$ M^{-1} and $K_2 = 130$ M^{-1}.

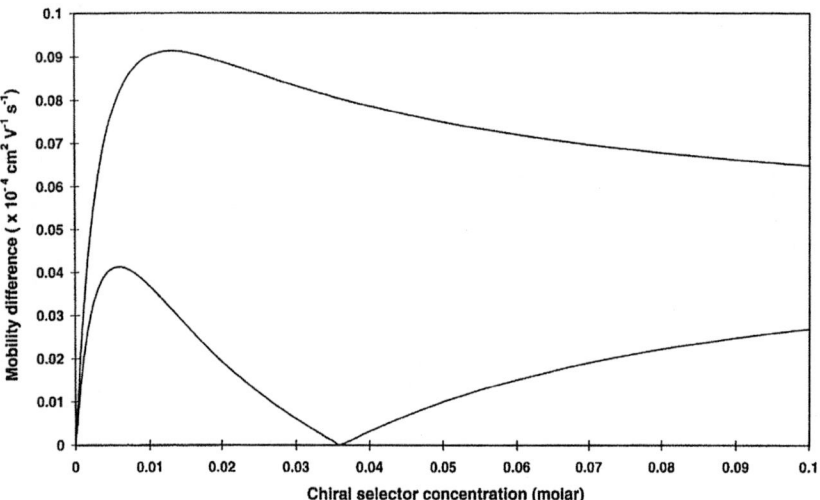

Figure 3.25. Mobility differences arising from the parameters given in Figures 3.23 and 3.24.

Because of the complexity of the selectivity equations it is simpler to examine their consequences graphically rather than to attempt to analyse them mathematically. In further work Vigh and co-workers have extended the selectivity model to describe the resolution between the enantiomers [24–27]. In these papers the influence of molecular diffusion, via the charge on the enantiomers is also considered. Because of the inevitable complexity of the resolution equation derived it is best to evaluate the consequences of certain parameters via computer modelling and the use of graphs to depict the response surfaces.

The influence of pH and β-cyclodextrin concentration on the separation of the enantiomers of the base dioxypromethazine has been considered by Ren and Liu [28]. These workers followed the approach of Vigh and co-workers in considering the complexation with cyclodextrins of both the charged and neutral forms of the enantiomers. Ren and Liu modelled the mobility difference of the enantiomers rather than selectivity. They simplified their model by assuming that the equilibrium constants for cyclodextrin complexation were the same for the neutral and protonated forms of the enantiomers ($K_{RCD} = K_{HRCD^+}$ and $K_{SCD} = K_{HSCD^+}$). The use of this rather unexpected approximation leads to an equation describing mobility difference which is similar to equation (3.5), but has the additional term $(1 + [OH^-]/K_b)$ in the denominator.

3.6 Conclusion

The separation and resolution of enantiomers using CE can be understood using various physical and mathematical models. Different models are appropriate according to the complexity of the analyte and selector and their interactions, and the degree of predictive accuracy required.

References

[1] Fanali, S.; Bocek, P. Enantiomer resolution by using capillary zone electrophoresis: Resolution of racemic tryptophan and determination of the enantiomeric composition of commercial epinephrine, *Electrophoresis* **1990**, *11*, 757–760.

[2] Fanali, S. Use of cyclodextrins in capillary zone electrophoresis. Resolution of terbutaline and propranolol enantiomers, *J. Chromatogr.* **1991**, *545*, 437–444.

Equation (3.15) is clearly complex but has been examined by Rawjee, Staerk, and Vigh who defined three types of enantiomers according to the selectivity of the chiral additive for the charged and neutral forms of the analyte.

Type I enantiomers are those where enantioselectivity is only seen for the uncharged forms ($K_{RCD^-} = K_{SCD^-}$ and $K_{HRCD} \neq K_{HSCD}$).

Type II enantiomers are those where enantioselectivity is only seen for the charged forms ($K_{RCD^-} = K_{SCD^-}$ and $K_{HRCD} \neq K_{HSCD}$).

And type III enantiomers are those where enantioselectivity is seen for both the neutral and the charged forms of the enantiomers ($K_{RCD^-} \neq K_{SCD^-}$ and $K_{HRCD} \neq K_{HSCD}$).

For Type I enantiomers it was assumed that the limiting mobilities of the complexes formed between the cyclodextrin and the charged enantiomers were equal, $\mu_{RCD^-} = \mu_{SCD^-}$. The first term in equation (3.15) therefore reduces to unity as $K_{RCD^-} = K_{SCD^-}$. The chiral selectivity therefore depends upon the size of the cyclodextrin complexation constants along with the pKa, pH, and cyclodextrin concentration.

For Type II and Type III enantiomers there is less simplification to equation (3.15) and the equations describing chiral selectivity remain rather complex. Again the chiral selectivity depends upon the size of the cyclodextrin complexation constants along with the pKa, pH, and cyclodextrin concentration.

[3] Guttman, A.; Paulus, A.; Cohen, A.S.; Grinberg, N.; Karger, B.L. Use of complexing agents for selective separation in high-performance capillary electrophoresis. Chiral resolution via cyclodextrins incorporated within polyacrylamide gel columns, *J. Chromatogr.* **1988**, *448*, 41–53.

[4] Nishi, H.; Fukuyama, T.; Terabe, S. Chiral separation by cyclodextrin-modified micellar electrokinetic chromatography, *J. Chromatogr.* **1991**, *553*, 503–516.

[5] Cole, R.O.; Sepaniak, M.J.; Hinze, W.L. Optimization of binaphthyl enantiomer separations by capillary zone electrophoresis using mobile phases containing bile salts and organic solvent, *J.H.R.C. & C.C.* **1990**, *13*, 579–582.

[6] Taylor, A.; Williams, D.A.R.; Wilson, I.D. Derivatised β-cyclodextrins combined with high field NMR for enantiomer analysis: application to ICI 185,282. *J. Pharm. & Biomed. anal.* **1991**, *9*, 493–496.

[7] Greatbanks, D.; Pickford, R. Cyclodextrins as chiral complexing agents in water, and their application to optical purity measurements. *Magnetic resonance in chemistry* **1987**, *25*, 208–215.

[8] Jorgenson, J.; Lukacs, K.D. Zone electrophoresis in open-tubular glass capillaries, *Anal. Chem.* **1981**, *53*, 1298–1302.

[9] Giddings, J.C. Generation of variance, "Theoretical Plates", Resolution and Peak Capacity in Electrophoresis and Sedimentation, *Sep. Sci.* **1969**, *4*, 181–189.

[10] Terabe, S.; Yashima, T.; Tanaka, N.; Araki, M. Separation of oxygen isotope benzoic acids by capillary zone electrophoresis based on isotope effects on the dissociation of the carboxyl group, *Anal. Chem.* **1988**, *60*, 1673–1677.

[11] Wren, S.A.C; Rowe, R.C. Theoretical aspects of chiral separation in capillary electrophoresis I. Initial evaluation of a model, *J. Chromatogr.* **1992**, *603*, 235–241.

[12] Terabe, S.; Isemura, T. Ion-exchange Electrokinetic Chromatography with Polymer Ions for the Separation of Isomeric Ions having Identical Electrophoretic Mobilities, *Anal. Chem.* **1990**, *62*, 650–652.

[13] Wren, S.A.C.; Rowe, R.C. Theoretical aspects of chiral separation in capillary electrophoresis III. Application to β-blockers, *J. Chromatogr.* **1993**, *635*, 113–118.

[14] Cruickshank, J.M. The clinical importance of cardioselectivity and lipophilicity in beta blockers, *Am. Heart Journal* **1980**, *100*, 160–178.

[15] Wren, S.A.C.; Rowe, R.C. Theoretical aspects of chiral separation in capillary electrophoresis II. The role of organic solvent, *J. Chromatogr.* **1992**, *609*, 363–367.

[16] Ferguson, P.D.; Goodall, D.M.; Loran, J.S. Systematic approach to the treatment of enantiomeric separations in capillary electrophoresis and liquid chromatography III. A binding constant-retention factor relationship and effects of acetonitrile on the chiral separation of tioconazole, *J. Chromatogr. A* **1996**, *745*, 25–35.

[17] Penn, S.G.; Liu, G.; Bergström, E.T.; Goodall, D.M.; Loran, J.S. Systematic approach to treatment of enantiomeric separations in capillary electrophoresis and liquid chromatography I. Initial evaluation using propranolol and dansylated amino acids, *J. Chromatogr. A* **1994**, *680*, 147–155.

[18] Wren, S.A.C.; Rowe, R.C.; Payne, R.S. A theoretical approach to chiral capillary electrophoresis with some practical applications, *Electrophoresis* **1994**, *15*, 774–778.

[19] Wren, S.A.C. Mobility measurements on dansylated amino acids., *J. Chromatogr. A* **1997**, *768*, 153–159.

[20] Wren, S.A.C. Chiral separation in capillary electrophoresis., *Electrophoresis* **1995**, *16*, 2127–2131.

[21] Sänger-van de Griend, C.E.; Gröningson, K.; Arvidsson, T. Enantiomeric separation of a tetrapeptide. Extension of the model for chiral capillary electrophoresis by complex formation of one enantiomer molecule with more than one chiral selector molecule, *J. Chromatogr. A* **1997**, *782*, 271–279.

[22] Rawjee, Y.Y.; Staerk, D.U.; Vigh, G. Capillary electrophoretic chiral separations with cyclodextrin additives. I. Acids: chiral selectivity as a function of β-cyclodextrin for fenoprofen and ibuprofen, *J. Chromatogr.* **1993**, *635*, 291–306.

[23] Rawjee, Y.Y.; Williams, R.L.; Vigh, G. Capillary electrophoretic chiral separations using β-cyclodextrin as resolving agent. II. Bases: chiral selectivity as a function of pH and the concentration of β-cyclodextrin, *J. Chromatogr.* **1993**, *652*, 233–245.

[24] Rawjee, Y.Y.; Vigh, G. A peak resolution model for the capillary electrophoretic separation of the enantiomers of weak acids with hydroxypropyl β-cyclodextrin containing background electrolytes, *Anal. Chem.* **1994**, *66*, 619–627.

[25] Rawjee, Y.Y.; Williams, R.L.; Vigh, G. Capillary electrophoretic chiral separations using cyclodextrin additives. III. Peak resolution surfaces for ibuprofen and homatropine as a function of the pH and the concentration of β-cyclodextrin, *J. Chromatogr. A* **1994**, *680*, 599–607.

[26] Rawjee, Y.Y.; Williams, R.L.; Buckingham, L.A.; Vigh, G. Effects of pH and hydroxypropyl β-cyclodextrin on peak resolution in the capillary electrophoretic separation of the enantiomers of weak bases, *J. Chromatogr. A* **1994**, *688*, 273–282.

[27] Williams, R.L.; Vigh, G. Buffer effects in the desionselective/ionoselective/ duoselective separation selectivity model-assisted optimization of the capillary electrophoretic separation of enantiomers I. Low-pH phosphate buffers, *J. Chromatogr. A* **1995**, *716*, 197–205.

[28] Ren, J.; Liu, H. Chiral separation of dioxypromethazine enantiomers by capillary electrophoresis using β-cyclodextrin as a chiral selector, *J. Chromatogr. A* **1996**, *732*, 175–181.

Method Development Strategies

4.1 The Purpose of the Method

The approach to method development will be partly determined by the purpose of the method and the constraints upon it. An important factor is whether the method is required for long term use or for short term problem solving. Methods which are intended for long term use must be designed with robustness and validation in mind. Another important consideration is the sample matrix. Samples of a pure drug substance in water for example are easier to analyse than those in complex biological matrices. Samples in matrices such as blood may require extraction and purification before analysis is possible. High ionic strength matrices in particular can be especially problematic.

A key factor for consideration is the approximate enantiomeric purity of the samples to be analysed. Samples in which both enantiomers are present in approximately equal amounts are significantly less demanding upon the method than those samples which have a high degree of enantiomeric purity. As most new pharmaceutical agents with an asymmetric centre are developed as a single enantiomers, the minor enantiomer may be regarded in the same light as any of the other related substances. Because of the high purity requirements for pharmaceutical agents the analytical methods must be able to quantify the minor enantiomer at the level of fractions of one percent.

The difficulties posed by samples with very different ratios of the two enantiomers are illustrated in the computer simulations shown in Figures 4.1 and 4.2. Figures 4.1 and 4.2 are idealised representations of CE peaks generated by using the assumption of normal probability distributions for each of the enantiomer peaks. The Figures cover variations in the level of the second enantiomer peak across two orders of magnitude, and at two different levels of selectivity. In each case the simu-

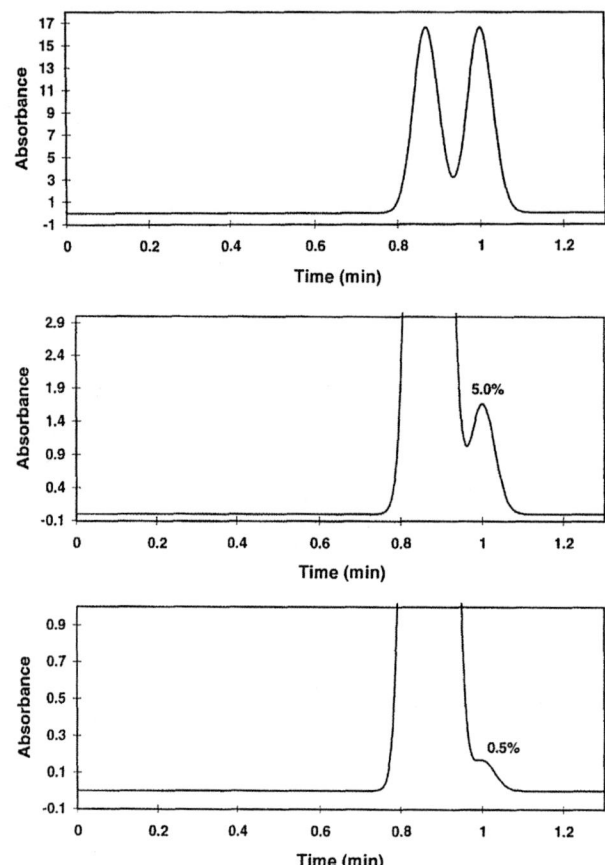

Figure 4.1. The separation of enantiomers obtained with a low selectivity separation. The minor enantiomer is present at 50%, 5%, and 0.5% of the total sample.

lations show cases where the two enantiomers are present in the ratios 50:50, 95:5, and 99.5:0.5. Figure 4.2 differs from Figure 4.1 in that the degree of separation between the enantiomers is twice as large. Whilst the degree of separation in Figure 4.1 is entirely adequate to give good quantification for a ratio of enantiomers of approximately 50:50, the uncertainty for a ratio of 95:5 would be significant and that for a ratio of 99.5:0.5 very considerable. For good quantification of the samples with the low levels of the minor enantiomer the degree of separation must be much higher than that depicted in Figure 4.1, for example that shown in Figure 4.2.

The representations shown in Figures 4.1 and 4.2 are idealised using perfectly symmetrical peak shapes. Such ideal behaviour is rarely seen in practice and usually the problem of peak tailing due to overloading or other problems means that the degree of separation required for good quantification of the minor enantiomer is even higher.

Low levels of the minor enantiomer place additional demands upon the analytical method. Unless the analyte has a very strong chromaphore it may be difficult to detect the minor enantiomer at a level of 0.5% or below. In order to achieve the required sensitivity for the minor enantio-

S-42

Chromatographia Supplement Vol. 54, 2001

Original

0009-5893/00/02   42-17 $ 03.00/0

mer large amounts of the sample may need to be injected onto the capillary. Large sample loadings can lead to the non symmetrical distortions of the main peak caused by electrophoretic dispersion [1]. In order to reduce the distortions to the main component peak high buffer concentrations may be required to try and compensate.

4.2 Factors Controlling Resolution

As has been discussed previously resolution between the enantiomer peaks depends upon two factors: the efficiency and the selectivity of the separation method. These two factors are incorporated in equation (4.1) which is used to describe resolution in CE [2].

$$R_S = \left(\frac{V}{32\,D}\right)^{0.5} \cdot \left(\frac{l}{L}\right)^{0.5} \cdot \frac{\Delta\mu_{ep}}{(\mu_{ep} + \mu_{eo})^{0.5}} \quad (4.1)$$

where V is the applied voltage, D the average diffusion coefficient for the two enantiomers, L is the total capillary length, l the length of the capillary from the inlet to the detector, $\Delta\mu_{ep}$ the electrophoretic mobility difference between the two enantiomers, μ_{ep} the average electrophoretic mobility of the two enantiomers, and μ_{eo} the electroosmotic mobility.

The first two terms of equation (4.1) are related to the efficiency of the separation system i.e. the extent to which the enantiomer peaks become dispersed during movement from the injector to the detector. The last term in equation (4.1) is related to the selectivity of the system i.e. the extent to which the peak maxima are separated.

4.2.1 Efficiency

The increase in resolution with the square root of the applied voltage that is expected from equation (4.1) has been demonstrated experimentally by Hutterer and Jorgenson [3]. The resolution between pairs of analytes was measured at voltages of 28 and 120 kV and the expected increase in resolution by a factor of 2.1 was in good agreement with that found experimentally.

Giddings has considered the ideal case in CE in which peak dispersion is caused by diffusion along the length of the capil-

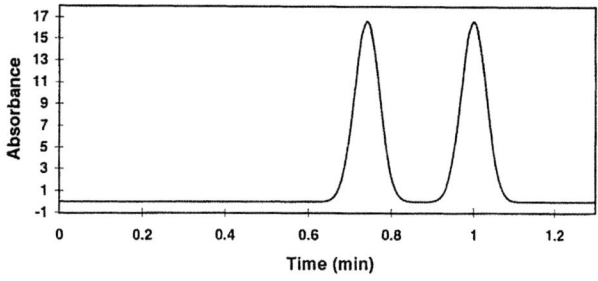

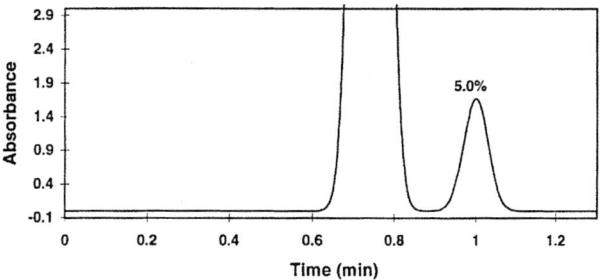

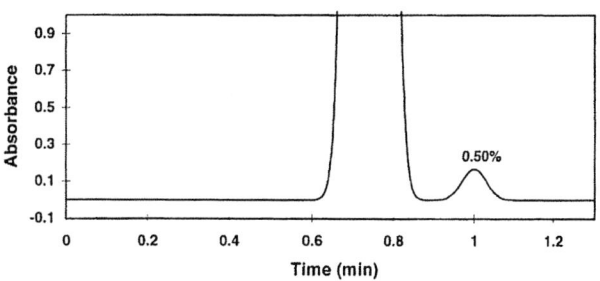

Figure 4.2. The separation of enantiomers obtained with a high selectivity separation. The minor enantiomer is present at 50%, 5%, and 0.5% of the total sample.

lary alone [4]. Under these conditions, and in the absence of electroosmotic flow, the efficiency can be related to both the voltage and the charge on the analyte as shown in equation (4.2).

$$N = \frac{FV_z}{2\,RT} \quad (4.2)$$

where N is the number of theoretical plates, F is the Faraday constant, V is the potential difference between the capillary inlet and the detector, z is the number of charges on the analyte, R is the gas constant, and T is the absolute temperature. At 298 K the value of $F/2RT$ is about 20 and so equation (4.3) may be employed as an approximation to equation (4.2).

$$N \approx 20\,V_z \quad (4.3)$$

Equation (4.2) has also been investigated by Kenndler and Schwer in experimental work on aromatic sulphonic and carboxylic acids [5]. Plate counts for mono and disulphonated benzene and naphthalene derivatives were recorded using potential drops of 8.84 and 4.42 kV. At the lower voltage the plate counts for the dis-

ulphonated analytes were greater than those for the monosulphonated analytes by a factor of 1.9. Doubling the potential drop lead to an increase in the plate count by an average of 1.95 for the monosubstituted analytes and an average of 1.75 for the disubstituted analytes. The measured plate counts were between 67 and 82% of the values expected from equation (4.2), with the agreement being closer for the monosubstituted analytes and with the smaller potential drop.

The differences between the theoretical and measured efficiency values was ascribed to differences in the conductivities between the analyte and sample zones, and to Joule heating. By using buffers of different pHs Kenndler and Schwer were also able to alter the degree of dissociation of three substituted benzoic acids. It was found that the measured plate count was directly proportional to the degree of dissociation (i.e. the partial charge on the analyte).

The results from other experiments can also be interpreted with the aid of equation (4.2). Figure 4.3 shows the separation

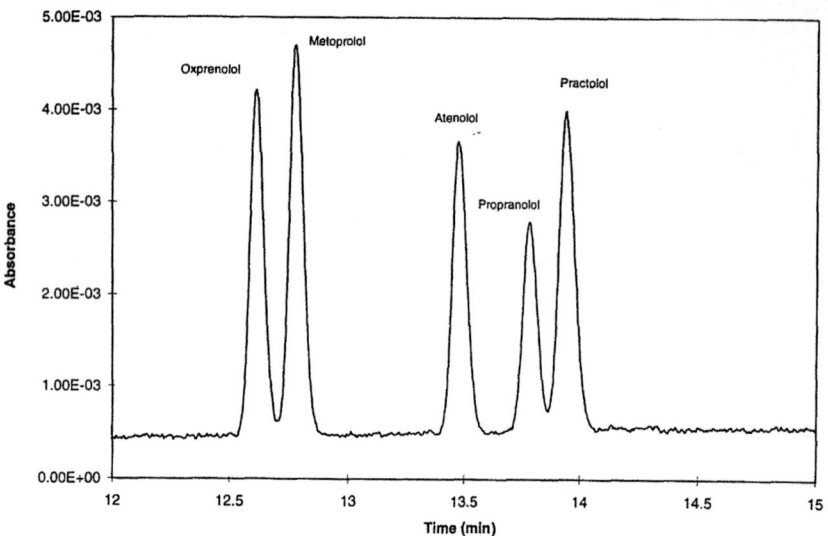

Figure 4.3. Separation of the β-blockers oxprenolol, metoprolol, atenolol, propranolol, and practolol in a 40 mM lithium phosphate buffer at pH 3.0.

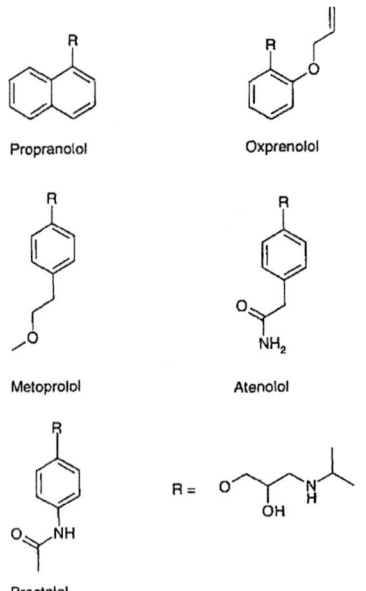

Figure 4.4. The structures of the β-blockers oxprenolol, metoprolol, atenolol, propranolol, and practolol.

of the β-blockers oxprenolol, metoprolol, atenolol, propranolol, and practolol at 298 K using a 40 mM lithium phosphate buffer at pH 3 and a potential drop of 17.5 kV. At pH 3 the β-blockers, whose structures are shown in Figure 4.4, all bear a single positive charge.

The β-blockers have a common side chain and very similar relative molecular masses, ranging from 259 for propranolol to 267 for metoprolol. Similar plate counts were obtained for all of the β-blockers ranging from 182,000 for practolol to 215,000 for oxprenolol. From equa-

tion (4.3) the maximum plate count that could be expected is 342,000 (assuming negligible electroosmotic flow). The similarity in the plate counts is consistent with the expectations from equation (4.2). The difference between the measured efficiency values and the theoretical maximum may again be due to thermal gradients caused by Joule heating.

As these β-blockers have the same charge and very similar masses their separation can be ascribed to small differences in shape along the lines discussed in Chapter 2.

In Chapter 2 the separation of some *n*-alkyl pyridines ranging in size from pyridine to hexyl pyridine was shown in Figure 2.2. The separation was carried out in a 40 mM lithium phosphate buffer at pH 2.5 and so all of the pyridines are fully charged. The voltage drop was 13.2 kV and so the expected maximum plate count is 256,000. The measured plate counts were found to be independent of the molecular weight of the analyte with the values for pyridine and 2-pentylpyridine for example being 220,000 and 190,000 respectively.

4.2.2 Selectivity

As has been discussed previously in Chapter 3 a key element in obtaining satisfactory resolution is the determination of buffer conditions which maximise the selectivity of the system. The selectivity is a measure of the extent to which the peak maxima are separated and this is related

to the last term in equation (4.1), which is given below in equation (4.4).

$$R_S \propto \frac{\Delta\mu_{ep}}{(\mu_{ep} + \mu_{eo})^{0.5}} \qquad (4.4)$$

where $\Delta\mu_{ep}$ is the difference in the electrophoretic mobilities of the two enantiomers, μ_{ep} is the average electrophoretic mobility, and μ_{eo} is the electroosmotic mobility.

An important element to consider is conditions under which the electrophoretic mobility difference is maximised. In Chapter 3 it was shown that for the simplest case, where the two enantiomer-chiral selector complexes have the same limiting mobilities, the mobility difference can be described by equation (4.5).

$$\Delta\mu_{ep} = \frac{[C](\mu_0 - \mu_1)(K_2 - K_1)}{1 + [C](K_1 + K_2) + K_1 K_2 [C]^2}$$
$$(4.5)$$

where [C] is the chiral selector concentration, K_1 and K_2 are the equilibrium constants for the formation of the two enantiomer-chiral selector complexes, and μ_0 and μ_1 are the electrophoretic mobilities of the free and complexed enantiomers respectively.

The electrophoretic mobility difference will therefore be increased by large differences between the equilibrium constants and large differences between the mobilities of the free and complexed enantiomers. In this simplest model of enantiomer separation there is an optimum chiral selector concentration which is inversely related to the affinity of the analytes for the chiral selector by equation (4.6).

$$[C]_{opt} = \frac{1}{\sqrt{K_1 K_2}} \qquad (4.6)$$

where $[C]_{opt}$ is the optimum chiral selector concentration and K_1 and K_2 are defined for equation (4.5).

The resolution also depends upon the numerator in equation (4.4). As both electrophoretic and electroosmotic mobilities are vector quantities, a consideration of both their size and direction is important. Under normal conditions the electroosmotic flow is towards the cathode and so the electroosmotic mobility has the same sign as the electrophoretic mobilities of cations. For cations, such as protonated bases, increasing the electroosmotic mobility will increase the size of the term $(\mu_{ep} + \mu_{eo})^{0.5}$ and so decrease the resolution. Basic enantiomers are often best analysed using coated capillaries or at low pHs where fused silica columns generate very low electroosmotic mobilities. There are exceptions to this general rule of thumb where the electrophoretic mobility difference is itself significantly altered by pH change. Exceptions can occur for example when the chiral selector shows different selectivities for the neutral and protonated forms of the enantiomers, and with chiral selectors such as macrocyclic antibiotic where the selectivity can also be a function of the pH.

For anions such as de-protonated acids the situation is more interesting as the electrophoretic and electroosmotic mobilities have opposite signs. Because of these sign differences it may be possible to manipulate conditions such that the term $(\mu_{ep} + \mu_{eo})^{0.5}$ is very small and so the resolution very large. In the extreme case, with the electrophoretic and electroosmotic mobilities having exactly the same magnitude but opposite sign, the resolution will be infinite but so will the analysis time. This close matching of mobilities has the effect of "stretching" a portion of the electropherogram and so increasing resolution. This stretching is obtained at the expense of longer analysis times and often "compression" of another portion of the electropherogram. The effect of separation stretch and compression can be illustrated by samples containing species with a wide range of electrophoretic mobilities. A good example is the resolution between the components of a dextran ladder. The ladder consists of glucose oligomers ranging in length from a single unit upwards. The oligomers were derivatised at their reducing ends with the fluorogenic agent Amino Pyrene Tri Sulphonic acid (APTS) to aid detection and provide the charge for the separation. The ladder sample was analysed both in the absence and presence of electroosmotic flow and the electropherograms are shown in Figure 4.5. Detection is achieved by exciting the analytes with an Argon ion laser at 488 nm and collecting the fluorescent signal at 520 nm.

Figure 4.5a shows the separation achieved using a 100 mM acetate buffer at pH 5.0 in a capillary which had been coated to reduce electroosmosis. The separation is caused by electrophoretic mobility differences alone with reversed polarity being employed (anode at the detector end). The ladder components are well resolved from each other up until long chain lengths.

Figure 4.5b shows the separation achieved using a 100 mM borate buffer at

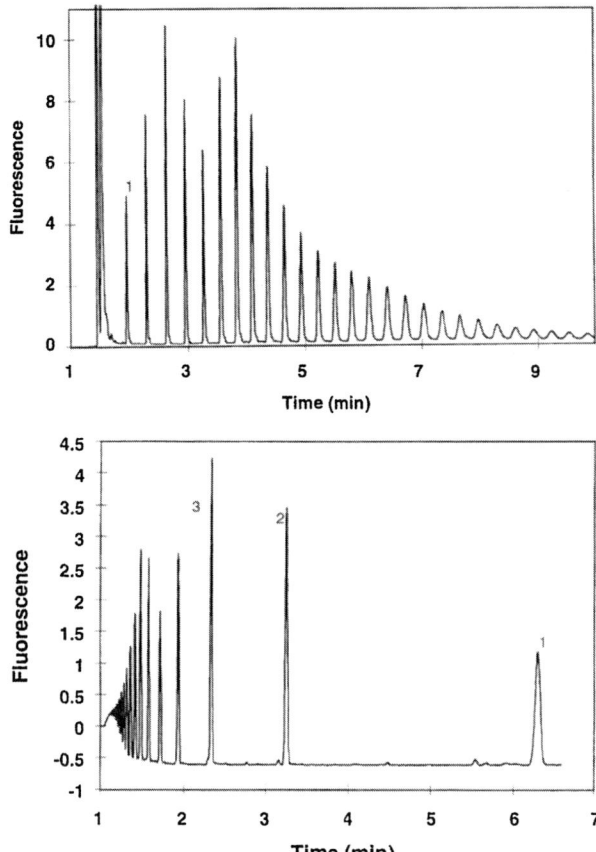

Figure 4.5. Separation of APTS derivatised glucose oligomers: (top) at pH 5 in a coated capillary, and (bottom) at pH 10 in a fused silica capillary.

pH 10.0 in a fused silica capillary. The separation is a reflection of the constant electroosmotic mobility and decreasing electrophoretic mobilities with increasing chain length. The migration order is the reverse of that seen in Figure 4.5a with the smallest analyte, glucose, having the smallest net mobility (value of $(\mu_{ep} + \mu_{eo})$). Figure 4.5b clearly shows how on one hand the separation space is stretched for the shorter oligomers, and on the other compressed for the longer oligomers.

Figure 4.6 shows how the resolution between oligomers of chain lengths n and $n+1$ (e.g. 3 and 4, or 4 and 5) varies with chain length in both the acetate and borate separation buffers.

With the acetate system and coated capillary (which gives a very low electroosmotic mobility) the resolution decreases only slightly with increasing chain length, and in a linear fashion. With the high electroosmotic mobility borate system the situation is more complex: with short chain lengths the resolution is very high but with longer chain lengths the resolution is low. In contrast with the acetate system the re-

solution in the borate system drops off rapidly with increasing chain lengths. With short oligomer chain lengths the electrophoretic mobility is high and so the term $(\mu_{ep} + \mu_{eo})$ is small; with the long chain oligomers however the electrophoretic mobility is low and so the term $(\mu_{ep} + \mu_{eo})$ is large.

The close matching of electrophoretic and electroosmotic mobilities can give very high resolution for a single pair of enantiomers but the same separation conditions will not be ideal for other pairs of enantiomers with significantly different electrophoretic mobilities. Best performance is obtained by tuning the separation conditions according to the properties of the analyte.

4.3 The Structure of the Analyte

In an ideal world our knowledge of enantiomer separation by CE would be such that information about the three dimensional structure of the analyte would be

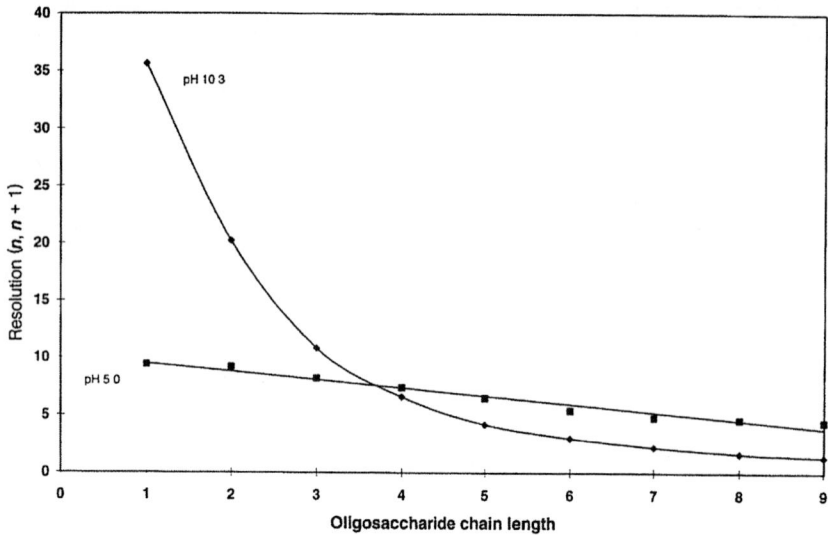

Figure 4.6. Resolution, as a function of chain length, between APTS derivatised glucose oligomers at pH 5 and 10.

sufficient to reliably predict the optimum separation conditions. At present this ideal world seems some way off in the future either near or distant. Whilst a great deal of useful data has been collected on a host of analytes using a wide range of chiral selectors, the information that has been extracted from the data is not sufficient for a complete understanding. There are however some clear trends and themes in enantiomer separation by CE and useful generalisations can be made. In addition the automated equipment and the simplicity of CE mean that experimentation is rapid.

A rational place to start in the development of a method for the separation of the enantiomers is knowledge of their structure. Some items for consideration are: the presence of functional groups capable of carrying a charge; the solubility and stability in water or polar organic solvents; the presence of a suitable chromaphore; and the proximity of the asymmetric centre to atoms or groups which can lead to stereospecific interactions.

A water soluble base which has an aromatic ring close to the asymmetric centre is a much better candidate for CE than a non-polar alkane which would probably be better handled by GC. Most enantioselective CE has been applied to analytes which have a reasonable solubility in aqueous buffer systems in the range between about pH 2 and 12. Moderate amounts of organic solvents such as methanol and acetonitrile have been used to obtain sufficient solubility for more hydrophobic compounds. More recently some workers have started to do CE using non-aqueous solvents such as dimethyl formamide and acetonitrile. This expansion in the range of solvents used increases both the range of compounds that can be analysed by CE and the range of binding interactions that can be exploited.

Analytes which can carry a charge are better candidates for CE because they will naturally have an electrophoretic mobility which can be manipulated. Neutral analytes can often be accommodated but will require a charged chiral selector or other competing additive which is itself charged. A commonly used competing additive which has a negative charge is Sodium Dodecyl Sulphate (SDS). SDS can compete with the cyclodextrin cavity for the hydrophobic parts of the analyte. This competition provides an electrophoretic mobility difference between the neutral analyte molecules which are bound to the chiral selector and those which are not. SDS also acts to solubilise relatively hydrophobic analytes. Analytes which do not posses a good UV chromophore will require either derivatisation or the use of a more complex detection approach such as inverse detection. An alternative approach is to use a different detection approach such as conductivity or mass spectrometry.

The separation of enantiomers requires both a difference in degrees of interaction with the chiral selector, and for that difference in interaction to be reflected in a difference in the electrophoretic mobilities. The difference in the degree of interaction with the chiral selector must arise from differences in one or more of the binding forces. The binding forces can be classified into groups such as dipole interactions, electrostatic forces, hydrophobic interactions, steric factors and so on.

Cyclodextrins are the most commonly used chiral selector in CE and they are normally thought to operate via an inclusion mechanism with a hydrophobic part of the analyte moving into the hydrophobic cyclodextrin cavity. It is thought that the two enantiomers interact differently or to different extents with the asymmetric centres around the cyclodextrin cavity leading to energy differences between the two inclusion complexes. If inclusion is to occur part of the analyte at least must have a hydrophobic character, for example an aromatic group. With very hydrophilic analytes there is much less chance of hydrophobic inclusion and a better approach might be to try and exploit enantioselective electrostatic interactions: for example by using chiral ion pair reagents in a relatively non-polar solvent. The skill in choosing and optimising the separation lies in using conditions which maximise the size of the stereoselective interactions and minimise the non-selective ones.

4.4 The Choice of Chiral Selector

The ideal chiral selector shows a high degree of enantioselectivity for a wide range of analyte structures, is very soluble in all common CE solvents, shows enantioselectivity which is independent of pH and other buffer parameters, is stable and UV transparent. The ideal selector should be commercially available in high purity, cheap, and show consistent performance from batch to batch. The selector should give rapid equilibrium kinetics with the analytes and should not lead to additional peak dispersion processes.

There are no chiral selectors for CE which fulfil all of these demanding requirements and so any particular choice inevitably involves compromise and a certain degree of experimentation. The choice of chiral selector is often based upon a consideration of which selectors have worked well for structurally similar analytes. More information on the selectors and the different classes of analytes is given in the following sections.

4.4.1 Analyte Variation

There are many examples of a particular chiral selector being applied effectively to a family of structurally related species. An effective approach to finding a suitable chiral selector may therefore be to consider which chiral selectors have worked for structurally related analytes.

It is also interesting to note how quite subtle structural changes to the analyte can result in significant differences in the degree of enantioselectivity observed. Sänger-van de Griend, Gröningsson and Westerlund used heptakis (2,6-di-O-methyl)-β-cyclodextrin (DM-β-CD) to separate the enantiomers of the local anaesthetics prilocaine, mepivacaine, ropivacaine and bupivacaine [6]. The electrophoretic mobility difference between the enantiomers was determined as a function of the cyclodextrin concentration for the anaesthetics and some structurally related analogues (Figure 4.7a and 4.7b).

The mobility difference between the enantiomers varied as a function of the DM-β-CD concentration and a maximum mobility difference was observed with each enantiomeric pair.

With ropivacaine and structural analogues (Figure 4.7a) the maximum mobility difference was in the order n-Bu > n-Pr > i-Bu > Me > i-Pr, with bupivacaine having a maximum mobility difference five and half times as large as the iso-propyl analogue. The mobility difference indicates that the selectivity is controlled by both the length and branching of the alkyl substituent. The data for prilocaine and analogues (Figure 4.7b) again shows the maximum mobility difference to depend on the nature of the alkyl side chain with the order being n-Pr > n-Bu = i-Bu > i-Pr. The maximum mobility difference between the enantiomers of prilocaine (n-Pr) was nearly three times as large as the maximum mobility difference for prilocaine's iso-propyl analogue.

Figure 4.8 shows the separation between the enantiomers of mepivacaine, ropivacaine, bupivacaine, and the ethyl and pentyl analogues using a buffer containing 10 mM DM-β-CD. The structural homologues are well separated from each other and their enantiomers well resolved.

Different enantioselectivities for structurally related analytes were observed by Rogan, Altria, and Goodall in their work on the β-stimulant salbutamol and some related substances which are shown in Figure 4.9 [7].

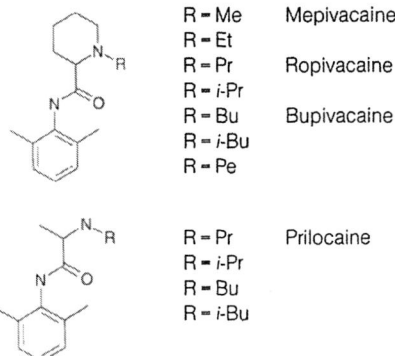

R = Me	Mepivacaine
R = Et	
R = Pr	Ropivacaine
R = i-Pr	
R = Bu	Bupivacaine
R = i-Bu	
R = Pe	

R = Pr	Prilocaine
R = i-Pr	
R = Bu	
R = i-Bu	

Figure 4.7. The structures of the local anaesthetics mepivacaine, ropivacaine, bupivacaine, and prilocaine and some structurally related analogues.

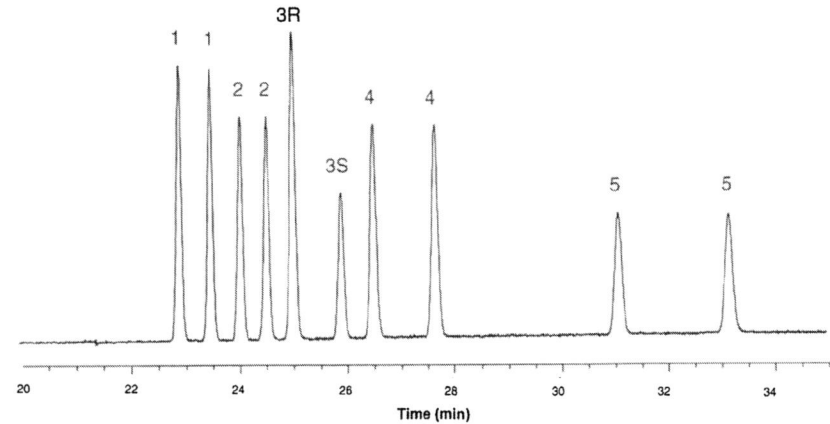

Figure 4.8. The separation between the enantiomers of mepivacaine (1), ropivacaine (3), bupivacaine(4), and the ethyl (2), and pentyl (5) analogues. Reproduced from [6] by kind permission.

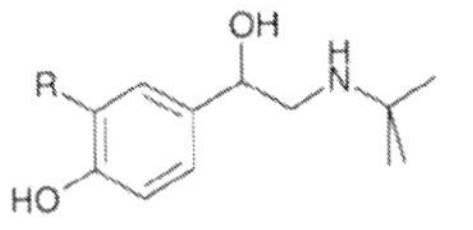

R = CH₂OH	Salbutamol
R = CHO	Aldehyde analogue
R = CH₃	Methyl analogue

Figure 4.9. The structures of salbutamol and the aldehyde and methyl related substances.

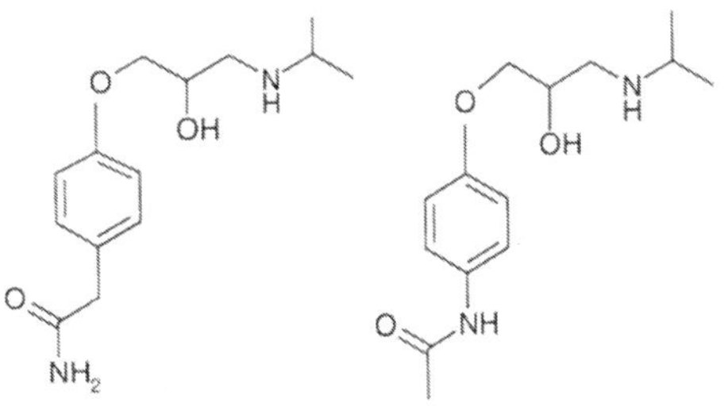

Atenolol   Practolol

Figure 4.10. The structures of the β-blockers atenolol and practolol.

The equilibrium constants for the formation of complexes between salbutamol and some of its related substances and DM-β-CD were determined in a citrate/phosphate buffer at pH 2.5. The enantioselectivities, as determined by the ratio of the equilibrium constants, were found to be 1.12 for salbutamol, 1.10 for the aldehyde analogue of salbutamol, and 1.20 for the deoxy analogue. Whilst some enantioselectivity was seen in each case, small changes to the substituents on the aromatic ring next to the asymmetric centre had an important influence on the degree of discrimination between enantiomers.

The small structural differences between the β-blockers atenolol and practolol (shown in Figure 4.10) also had a signif-

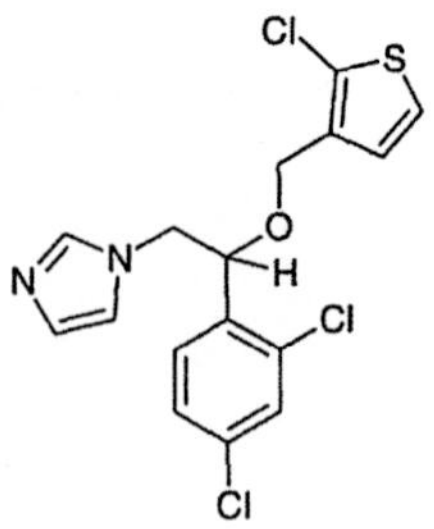

Figure 4.11. The anti-fungal agent tioconazole.

icant impact on the enantioselectivity seen with DM-β-CD [8]. The atenolol enantiomers gave a maximum mobility difference of 0.013×10^{-4} cm^2 V^{-1} s^{-1} at a DM-β-CD concentration of 30 mM, whilst the practolol enantiomers gave a maximum mobility difference of 0.015×10^{-4} cm^2 V^{-1} s^{-1} at concentration of 20 mM. These differences in both affinity and enantioselectivity are seen even though the β-blockers are isomers of each other, and the minor structural differences are a long way from the asymmetric centre.

Different enantioselectivities with structurally similar analytes have also been observed with macrocyclic chiral selectors such as the antibiotic rifamycin B [9]. A 25 mM solution of the antibiotic was used to measure the resolution between the enantiomers of some β-blockers and other amino alcohols such as ephedrine and epinephrine. The degree of resolution varied from 0.4 for oxprenolol to 3.1 for terbutaline. In general the resolution was better when the alcohol group was next to the aromatic ring, and with secondary rather than primary amines.

4.4.2 Chiral Selector Variation

Variation in the chiral selector chosen can have a major impact on the degree of enantioselectivity, even when the differences between the chiral selectors appears at first appearance to be minor.

Fillet and co-workers examined the enantioselectivity for the β-blocker propranolol of a series of β-cyclodextrins in the concentration range 1–50 mM [10]. The resolution obtained with β-cyclodextrin was compared with that obtained from the methyl, di-methyl, tri-methyl, hydroxypropyl, carboxymethyl, carboxyethyl, and succinyl derivatives. The resolution was shown to be a function of the concentration of the cyclodextrin with the different cyclodextrins giving different maximum resolutions at different concentra-

tions. The carboxyethyl cyclodextrin gave a resolution of 1.6 at a concentration of 15 mM whereas the carboxymethyl derivative gave a resolution of 4.4 at 10 mM. The parent β-cyclodextrin gave a resolution of 1.1 at a concentration of 10 mM whereas the methyl derivative gave a resolution of 1.9 at 5 mM. These data suggest that the different cyclodextrins give differences in both the degree of absolute, and the degree of enantioselective, binding for the propranolol enantiomers.

The binding constants of β-cyclodextrin and its methyl, dimethyl and hydroxypropyl analogues to the enantiomers of the anti-fungal agent tioconazole (Figure 4.11) have been determined by Penn and co-workers [11].

Again the different cyclodextrins were shown to give different degrees of absolute and stereoselective binding for the tioconazole enantiomers. The binding constants increase in the order β-cyclodextrin < methyl analogue < dimethyl analogue. The best enantioselectivity was found with the β-cyclodextrin parent. Three hydroxypropyl derivatives of β-cyclodextrin which differed in their average degree of substitution were also studied. With the hydroxypropyl derivatives the binding constants increase in the order: DS 1.0 < DS 0.6 < DS 0.9 with enantioselectivities DS 0.6 < DS 1.0 ≈ DS 0.9. As most substituted cyclodextrins contain a range of species with differing degrees of substitution the results with tioconazole illustrate the need for batch to batch consistency with the chiral selector.

Most analysts have chosen to use chiral selectors which are commercially available and well characterised such as cyclodextrins. These selectors are then evaluated sequentially until a satisfactory enantiomeric separation has been achieved.

A recent interesting departure from this procedure is the combinatorial synthesis approach of Chiari and co-workers [12]. Libraries of cyclic hexapeptides were prepared and, following fractionation based upon their hydrophobicities, used to separate some amino acids which had been derivatised with dinitrophenylhydrazine. The use of the library approach gave a more efficient determination of which amino acids were needed in which position to achieve the best and broadest range of enantioselectivity.

4.4.3 Screening Selector Type and Concentration

The analysis in the previous sections has shown how the performance of a given chiral selector can vary significantly even for analytes which, superficially at least, are very similar structurally. In addition for a given analyte the enantioselectivities shown by different related selectors can also be significantly different. Because of the uncertainties around structure-enantioselectivity relationships there is a need to screen different chiral selectors as part of method development. The observation that there are optimum chiral selector concentrations, and that these can vary across several orders of magnitude depending upon the analyte and chiral selector, means that a wide range of concentration ranges should be screened.

An example of such a screening approach is the "cyclodextrin array chiral analysis" advocated by Guttman and others [13–15]. This approach employs a 50 μm × 27 cm coated capillary (to reduce electroosmotic mobility) and either a pH 2.5 buffer for basic analytes, or at pH 8.0 buffer for acidic analytes. A series of four neutral cyclodextrins are then employed at two different concentrations in order to find the most promising starting conditions for optimisation work. Beta cyclodextrin is used at 3 and 15 mM, γ-cyclodextrin at 10 and 50 mM, hydroxypropyl β-cyclodextrin at 10 and 100 mM, and dimethyl β-cyclodextrin at 10 and 50 mM. Following the identification of the best cyclodextrin, additional experiments can be performed to optimise the concentration. If no resolution is obtained following the initial screening experiments the authors suggest employing a buffer of the same pH as the pK_a of the analyte. The authors list the best conditions obtained by the application of the screening approach to the enantioseparation of alprenolol, propranolol, metaproterenol, methoxyphenamine, terbutaline, homatropine, fenoprofen, naproxen, ibuprofen, and the dansyl derivatives of phenylalanine, leucine, aspartic acid and glutamic acid.

The array screening approach is an efficient one although there is the possibility that the low concentrations employed are still too high for analytes with very high affinities for the cyclodextrins used. For situations where the equilibrium constants are greater than 1000 M^{-1} the optimum selector concentration will be below 1 mM. The possibility of overshooting the optimum concentration can be guarded against by injecting the analyte in the absence of any cyclodextrin and comparing migration times. For cases where the selector and analyte have a very large affinity there will be a large change in the observed electrophoretic mobility even upon the use of low concentrations of the selector.

4.5 Other Separation Conditions

4.5.1 The Analyte

As most single enantiomer pharmaceutical agents are of very high optical purity the analytical samples will contain one enantiomer in a large excess. Such sample are significantly harder to analyse than racemic mixes. Racemic mixtures, or samples which have large levels of the minor enantiomer, are to be preferred for the early stages of method development. With racemic mixtures it is straightforward to determine when the enantiomers have been separated as two equal peaks will be seen. With an enantiomerically pure material the small peak which is pulled out from the main component may not be the minor enantiomer but could be another related substance. With racemic mixtures both enantiomers are easy to detect and so small amounts of the analyte can be injected so avoiding overloading symptoms such as tailing. If possible the sample should be dissolved in a solution which has the same or a lower ionic strength than that of the buffer.

4.5.2 Choice of Capillary

Short, narrow capillaries give fast analysis times but give less sensitivity and sample capacity than longer and wider capillaries. The shorter and narrower capillaries are therefore more appropriate for screening chiral selectors and method optimisation. Longer and wider capillaries may be required for the routine analysis of more challenging samples such as those of high enantiomeric purity.

The chiral selector screening experiments can be performed quickly by using short narrow capillaries to minimise the analysis times. In an extension of this approach Aumatell and Guttman used the short section of the capillary from what is normally the outlet to the detector to per-

form the separation [16]. In this method the effective length of the capillary was reduced giving faster analysis although at the cost of reduced resolution. The resolution is reduced as the separation which would continue to develop beyond the detector is not seen (see equation (4.1)). Such approaches allow faster screening but carry a greater risk that a promising selector is overlooked as no measurable resolution is seen.

Both simple bare fused silica and coated capillaries are appropriate. Bare silica has the advantage of cost and availability in a wide range of sizes. Coated capillaries are designed to reduce analyte interactions and provide a more stable electroosmotic flow over a wider pH range than fused silica. Coatings designed to eliminate or reduce the size of the flow can lead to higher resolution as can be seen from equation (4.1).

4.5.3 Buffer pH

After the choice of the type and concentration of the chiral selector the buffer conditions are the most important factors in determining the success of enantiomer separations. The key points are the pH chosen, the type of buffer and the concentration.

The pH employed is primarily determined by the properties of the analyte such as the pK_a value. With a neutral chiral selector the analyte must be charged and this leads to the general use of low pHs for bases and neutral to high pHs for acids. For the simplest separation cases where the enantioselectivity of the neutral form is the same as, or lower than, that of the charged form of the analyte the main pH consideration is connected with electroosmotic mobility. For fused silica capillaries the electroosmotic mobility increases with pH. From equation (4.1) we see that for bases reducing electroosmotic mobility will improve resolution and so operation at a low pH is desirable. For acids the resolution will be low when μ_{eo} is much greater than $-\mu_{ep}$, but can be very high when μ_{eo} is approximately equal to $-\mu_{ep}$.

For some analytes the enantioselectivity is higher with the neutral rather than the charged forms of the enantiomers. With these analytes the pH employed is a compromise between maximising enantioselectivity and maintaining some electrophoretic mobility. Using a pH equiva-

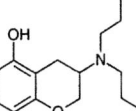

Figure 4.12. The structure of 5-hydroxy-3-(di-*n*-propylamino) chroman.

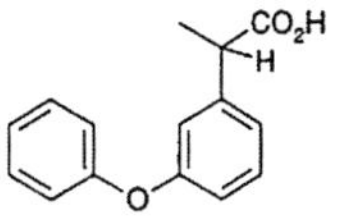

Figure 4.13. The structure of fenoprofen.

lent to the analyte pK_a value ensures equal proportions of the charged and neutral forms of the analyte. The situation of different enantioselectivities for neutral and charged forms of the analyte is discussed in more detail in Chapter 3. For a comprehensive theoretical treatment of the possible changes to enantioselectivity as a function of both pH and chiral selector concentration the reader is referred to the work of Vigh and co-workers e.g. [17]. An interesting example is that given by Baumy and others of β-cyclodextrin and 5-hydroxy-3-(di-*n*-propylamino) chroman (5-OH-DPAC) [18]. The analyte, Figure 4.12, was analysed using 6 mM β-cyclodextrin in a 50 mM phosphate-borate buffer across the pH range 4.5–12. Two distinct maxima were observed centred on pH 7 and pH 11.75 respectively. The two maxima are as a result of the β-cyclodextrin interacting with the cationic and anionic forms of the analyte.

In certain cases the pH might also be used to produce a charge difference between the two enantiomer-selector complexes. In the absence of a selector the enantiomers have the same pK_a values. The two transient enantiomer-selector complexes are however diastereoisomers and so their pK_a values may be shifted to different extents. In Chapter 2 we saw that for isomers such as the methylpyridines the charge difference was maximised at a pH equal to the average of the two pK_a values. As the formation of the complexes seems unlikely to alter the pK_a by much, using a pH equal to the analyte pK_a value is a sensible strategy. An example of fine tuning the pH is that of fenoprofen given by Guttman et al. [13]. The enantiomers of fenoprofen (Figure 4.13) were resolved using a buffer with a pH of 4.66 after a buffer with a pH of 8.0 had failed.

Table 4.I. Commonly used buffers for CE

pKa value (25 °C)	Species
2.15	Phosphoric acid (pK₁)
3.13	Citric acid (pK₁)
3.75	Methanoic acid
4.75	Ethanoic acid
4.75	Citric acid (pK₂)
6.10	MES (2-(N-Morpholino) ethanesulphonic acid)
6.40	Citric acid (pK₃)
7.20	Phosphoric acid (pK₂)
7.76	Triethanolamine
8.06	TRIS (Tris(hydroxy- methyl)aminomethane)
9.23	Boric acid
10.7	Triethylamine
12.3	Phosphoric acid (pK₃)

(Note: pK subscripts: pK_1, pK_2, pK_3)

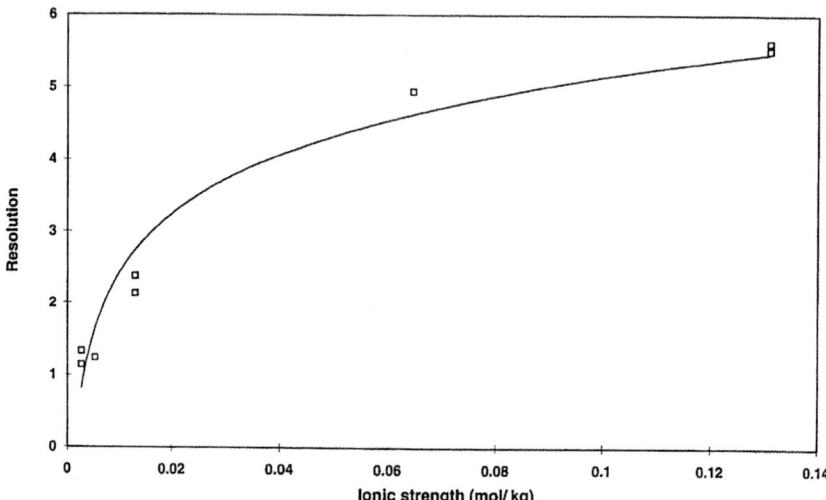

Figure 4.14. The structures of clenbuterol and picumeterol.

Figure 4.15. The increase in the resolution between the enantiomers of clenbuterol with increasing buffer ionic strength. Figure adapted from [24] with permission.

4.5.4 The Choice of Buffer and Buffer Concentration

The buffer chosen will depend on the pH chosen for the separation and other factors such as compatibility with the analyte and the detection system. The maximum buffering capacity is obtained at a pH equal to the pK_a of the buffer. The buffering capacity reduces the further the pH is away from the pK_a value and so a general recommendation is to restrict the use of a buffer to less than one pH unit either side of the pK_a value. The buffer should not interfere with the detection system and so UV transparency or compatibility with MS are important. There are sometimes conflicts between the desirable properties of the buffer used. For example whilst phosphate buffers are very useful with UV detection they are not ideal for MS because of their lack of volatility. Buffers based upon acetic or formic acids (ethanoic and methanoic acids) have the volatility required for MS detection but absorb significantly at low UV wavelengths.

The buffer determines the current generated and so low conductivity and zwitterionic buffers are popular. A list of some of the acids and bases which are commonly used to prepare CE buffers is given in Table 4.I. A much more comprehensive list of acids and bases and their pK_a values can be found in reference [19].

Good buffering capacity is also important as electrolysis of the buffer solution means that the pH will tend to decrease at the anode and increase at the cathode [20]. The electrolysis can have consequences such as poor reproducibility of migration times due to variations in the buffer pH. These problems can be minimised by en-suring good buffering capacity, changing the buffer regularly, using large volume electrode vials [21], and ensuring that the capillary and electrodes are well separated [22].

A common problem with the determination of low levels of the minor enantiomer in an optically pure sample is that of electrophoretic dispersion. A large sample mass has to be injected in order to see the minor enantiomer and this causes significant fronting or tailing to the peak of the major enantiomer. The tailing or fronting can obscure other components, such as the minor enantiomer, which have similar electrophoretic mobilities to that of the main component. The problem of electrophoretic dispersion was studied by Mikkers and co-workers [23]. The solution requires either a buffer concentration of about 100 times that of the sample, or a close match in electrophoretic mobilities between the analyte and the buffer component of the same charge.

Altria and others [24] studied the impact of changing the buffer ionic strength on the resolution between the enantiomers of clenbuterol and those of picumeterol, an amino alcohol development compound (Figure 4.14).

The buffer of pH 4.0 was prepared from 0.1 M citric acid and 0.2 M disodium phosphate and also contained 16 mM β-cyclodextrin. The use of ionic strengths from about 0.004 up to 0.13 mol kg⁻¹ lead to a significant change in the amount of peak tailing with much better peak symmetries at high ionic strengths. The reduction in tailing at higher ionic strengths gave an increase in resolution between the enantiomers of clenbuterol by about a factor of five as seen in Figure 4.15.

Improvements in resolution with increasing buffer concentration were also seen by Baumy et. al. [18]. They observed than the measured efficiency of the enantiomers of 5-OH-DPAC (Figure 4.12) increased by 450% upon increasing the buffer concentration from 10 mM to 50 mM. The resolution increased approximately in line with the square root of the increase in efficiency, as expected from theory.

Rickard and Bopp [25] studied the affect of varying both sample loading and buffer concentration on the tailing, and the resolution between enantiomers of the development compound LY248686 shown in Figure 4.16.

With a constant mass of sample injected the tailing decreased from 2.24 with

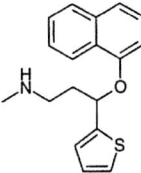

Figure 4.16. Development compound LY 248686.

a 10 mM Tris-phosphate buffer, to 1.61 with a 200 mM buffer. The reduction in tailing lead to an increase in the resolution from 2.17 to 4.17. In other experiments the buffer concentration was kept constant whilst the mass injected was varied by either increasing the sample concentration or injection volume. At low sample masses the degree of tailing and resolution did not change with increasing sample mass, but at higher sample masses tailing increased and resolution dropped. In Figure 4.17 the resolution between the enantiomers is shown as a function of the ratio of the mass of sample injected (in nanograms) to the buffer concentration (mM). The two data sets show: a) the data obtained from using a constant sample mass with variable buffer concentration, and b) the data from a variable mass of sample with a constant buffer concentration. Whilst Figure 4.17 show some scatter at very low ratios of sample mass/buffer concentration because of the imprecision in measuring very small peaks, the trend is clear. The data obtained from the two approaches are very similar. Unless the mass of the sample is very low relative to the buffer concentration, the resolution will decrease as the mass of sample injected increases.

Electrophoretic dispersion may also be reduced by a good match between the electrophoretic mobilities of the analyte and the buffer ion carrying the same charge. With many common buffer ions mobility matching is difficult to achieve as the analyte ions tend to be much larger than the buffer ions and so have smaller electrophoretic mobilities. The electrophoretic mobility of some buffer ions can however be adjusted by altering the degree of dissociation and so their effective mobilities. This approach has been used most widely with weak acid buffer co-ions.

Sustacek, Foret and Bocek [26] examined different buffer co-ions to analyse a mixture of 3-Phenyllactic acid (PL), Phenylacetic acid (PA), and the 2-, 3-, and 4-Hydroxyphenylacetic acids (2-HPA, 3-HPA, and 4-HPA). Buffer system at pH 4

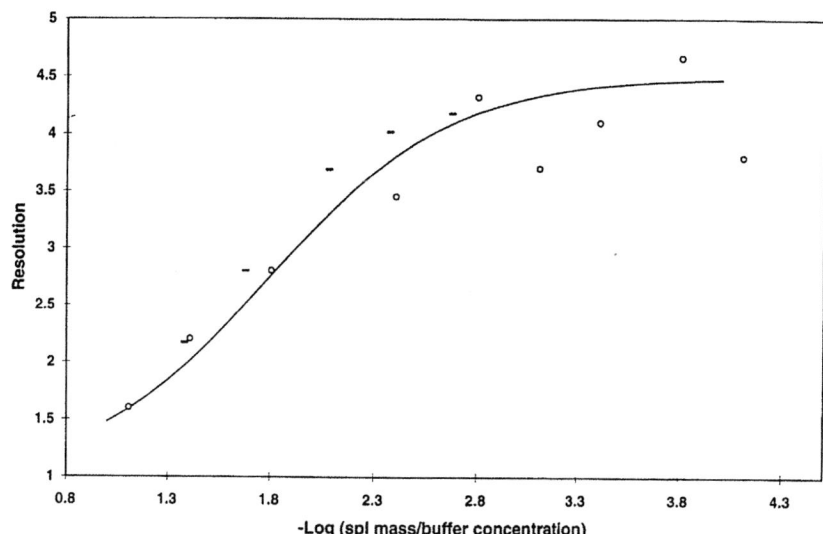

Figure 4.17. The resolution between enantiomers of development compound LY 248686 as a function of the ratio of the amount of sample injected to the buffer concentration.

Table 4.II. Ionic mobilities and effective mobilities at pH 4.0 for buffer ions and analytes.

Component	pK_a	Ionic mobility $(10^{-5}\,cm^2\,V^{-1}\,s^{-1})$	Effective mobility $(10^{-5}\,cm^2\,V^{-1}\,s^{-1})$
Trichloroacetic acid	0.64	36.2	36.2
2-Hydroxyisobutyric acid	3.97	33.5	17.3
Glutamic acid	4.32	27.0	8.7
4-Aminobutyric acid	4.0	≈ 34	≈ 17
3-Phenyllactic acid	–	≈ 25	17.4
Phenylacetic acid	–	≈ 30	10.4
2-Hydroxyphenylacetic acid	–	≈ 30	8.9
3-Hydroxyphenylacetic acid	–	≈ 30	9.9
4-Hydroxyphenylacetic acid	–	≈ 30	8.2

were prepared using 4-aminobutyric acid as the counter ion and either: a) 10 mM trichloroacetic acid, b) 100 mM trichloroacetic acid, c) 20 mM 2-hydroxyisobutyric acid, or d) 30 mM glutamic acid. At pH 4.0 the effective mobility of 2-hydroxybutyric acid is very similar to that of 3-phenyllactic acid and the effective mobility of glutamic acid is similar to those of the phenylacetic acids (Table 4.II).

The separations produced are shown in Figure 4.18. With the 10 mM trichloroacetate buffer all the peaks tailed badly and so resolution was very poor. With the 100 mM trichloroacetate buffer the tailing is greatly reduced but is still significant. The analysis time is long as a smaller voltage had to be used to reduce the current generated. With the 20 mM 2-hydroxybutyric acid buffer the phenyllactic acid peak is very sharp whilst the other analytes tail badly. With the 30 mM glutamic acid buffer the hydroxyphenylacetic acids give sharp peaks whilst the phenyllactic acid and phenylacetic acid peaks front (they have *higher* mobilities than that of the glu-

tamate). Close inspection shows that 2-HPA (which has an effective mobility closest to that of glutamate) has the most symmetrical peak and that 3-HPA and 4-HPA front and tail respectively.

Mobility matching has also been used to improve the resolution between the enantiomers of basic analytes. Stålberg and co-workers used hydroxypropyl-β-cyclodextrin (HPBCD) in buffers prepared from tetraalykylammonium ions and phosphoric acid to resolve the enantiomers of phenyramidol and sotalol (Figure 4.19) [27]. With a buffer containing 10 mM of tetrabutylammonium (TBA) phosphate, tailing of the *l*-sotalol peak meant that the peak due to a 5% level of the *d* enantiomer was poorly resolved. Changing the buffer co-ion to a similar concentration of the tetrapentylammonium salt resulted in a major improvement to peak shape and baseline resolution of the enantiomers.

In situations of sample overloading, mis-matching between analyte and buffer ion mobilities can also confuse the optimi-

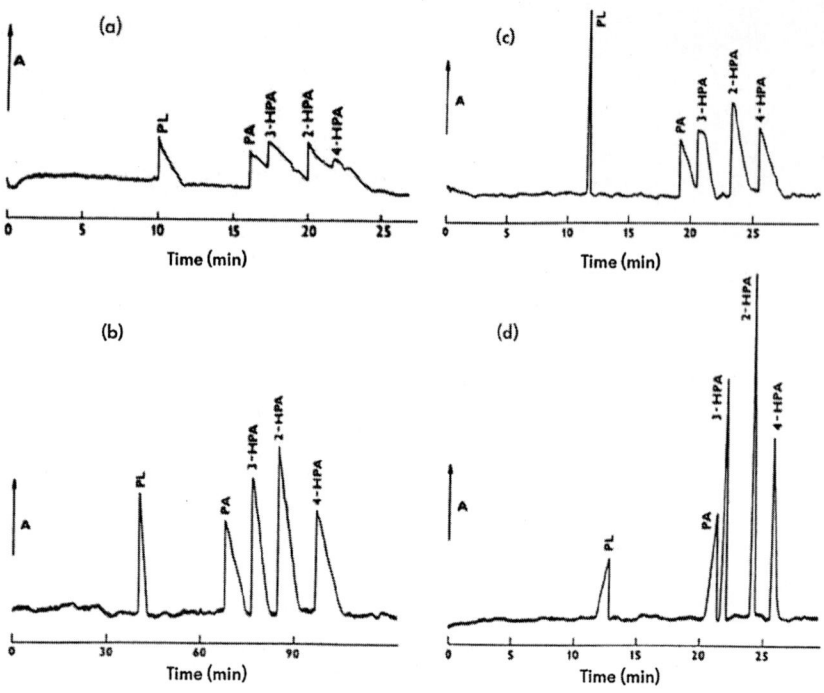

Figure 4.18. The peak shapes of some aromatic carboxylic acids using buffer co ions of differing strengths and electrophoretic mobilities. a) 0.01 M trichloroacetate, b) 0.1 M trichloroacetate, c) 0.02 M hydroxybutyrate, d) 0.03 M glutamate. Reprinted from [26] with permission from Elsevier Science.

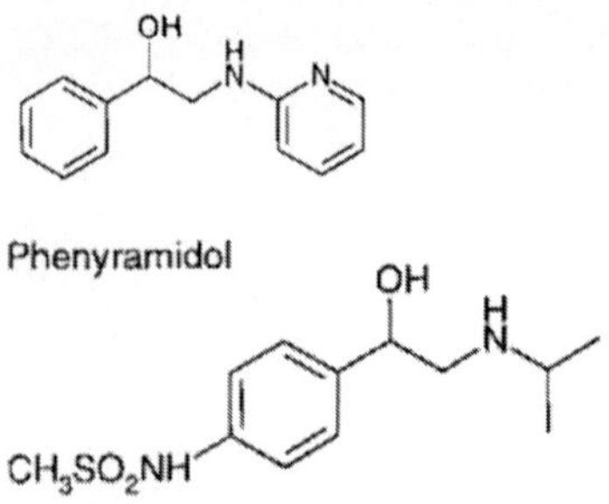

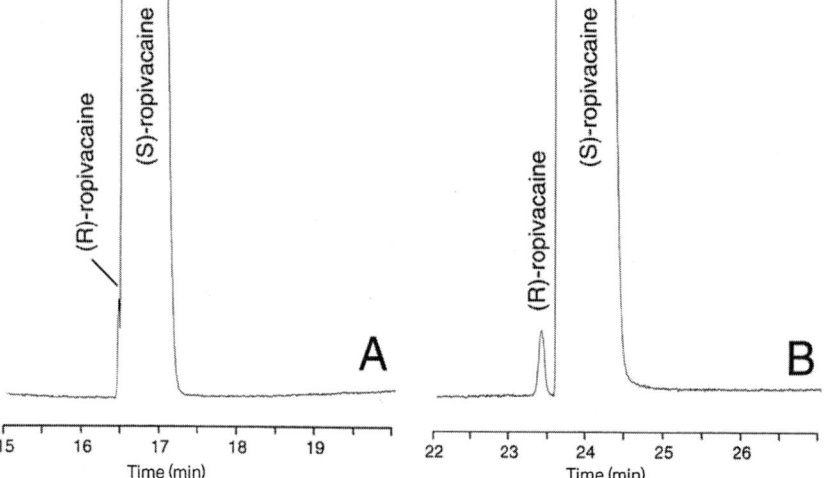

Figure 4.20. Electropherograms of a solution containing 0.1% (*R*)-ropivacaine in (*S*)-ropivacaine in buffers prepared using phophoric acid and either A) sodium hydroxide or B) triethanolamine. Reproduced from [6] with permission.

sation of the chiral selector concentration. Changing the selector concentration also changes the effective mobilities of the enantiomers of the analytes, as they will spend more time in the complexed form. With phenyramidol in 10 mM of TBA phosphate the enantiomer peaks fronted with 5 mM HPBCD reducing resolution. Increasing the HPBCD concentration to 20 mM resulted in lower effective mobilities for the enantiomers and so more symmetrical peaks and higher resolution. In

any approach to improving resolution it is important to ensure that any changes which are made to improve the selectivity do not inadvertently reduce the separation efficiency.

The mobility matching principle has been demonstrated by Williams and Vigh who used a series of tetraalkylammonium phosphates as the buffers for the separation of the enantiomers of some (α-hydroxymethylbenzyl)trialkyl ammonium ions [28]. The separation buffers were prepared

Phenyramidol

CH₃SO₂NH

Figure 4.19. The compounds phenyramidol and sotalol.

from 15 mM β-cyclodextrin and 50 mM phosphoric acid and were titrated to pH 2.2 using either tetraethylammonium hydroxide, tetrapropylammonium hydroxide, or tetrabutylammonium hydroxide. The buffers had the same ionic strength, buffering capacity and cyclodextrin concentration. With the tetraethylammonium buffer all of the analyte peaks fronted as their electrophoretic mobilities are lower than that of the tetraethylammonium ion. With the tetrapropylammonium buffer good peak symmetries were obtained for the enantiomers of the (α-hydroxymethylbenzyl)trimethyl, and triethyl ammonium ions whilst the less mobile analyte ions tailed. With the tetrabutylammonium buffer the more mobile analyte peaks fronted whilst those with lower mobilities now gave good peak shapes.

The mobility matching approach has also been employed by Williams and Vigh for anionic enantiomers [29]. Oligomeric anionic co-ions with electrophoretic mobilities in the range $5-45 \times 10^{-5}$ cm² V⁻¹ s⁻¹ were prepared by reacting poly ethylene glycol mono methyl ethers of different chain lengths with chlororsulphonic acid. The enantiomers of 3,5-dinitrobenzamido phenylalanine were separated in systems containing 8 mM β-cyclodextrin and 50 mM ε-aminocaproic acid at pH 4.4, with either 25 mM of chloride or 25 mM of one of the oligomeric sulphates as the co-ion. The analyte concentration was 10% of that of the buffer co-ion. In the chloride system the enantiomers gave tailing peaks but with the oligomeric sulphate co-ion symmetrical peaks and higher plate counts were obtained. The enantioselectivity was not altered by the change of buffer co-ion.

The buffer ions can also alter the degree of resolution between enantiomers because of their influence on electroosmotic flow. Figure 4.20 shows the electropherograms produced by a sample containing 0.1% of (*R*) ropivacaine in (*S*) ropivacaine using two different buffer cations [6].

In both cases the buffers contained 10 mM DMBCD and were prepared from 100 mM phosphoric acid adjusted to pH 3.0. In Figure 4.20a the pH was adjusted with sodium hydroxide and in Figure 4.20b the pH was adjusted with triethanolamine. The triethanolamine leads to an increase in analysis time and a significant improvement in resolution. It is thought that triethanolamine both reduces electrodispersion and also coats the capillary wall leading to a reversal in the direction of the electroosmotic flow.

4.5.5 Electrical Field Strength

The size of the applied voltage has a significant impact upon the resolution. Guttman and Cooke studied the resolution between the enantiomers of propranolol using hydroxypropyl-β-cyclodextrin [30]. The resolution varied with the field strength as seen in Figure 4.21.

Initially the resolution increases with increasing field strength due to improvements in the separation efficiency, as expected from equations (4.1) and (4.2). At higher field strengths the resolution reaches a maximum value and then declines, probably because the joule heating causes significant thermal gradients and so additional band broadening. Similar resolution optima with increasing field strength were observed for the enantiomers of a range of racemic drugs by Nielen [31].

4.5.6 Operating Temperature

The resolution is also altered by the temperature of the buffer system employed. Guttman and Cooke [30] showed a decrease in the resolution between enantiomers of propranolol as the temperature was increased from 20 °C to 50 °C as seen in Figure 4.22.

Similar decreases in resolution with increasing temperature were observed for the enantiomers of both clenbuterol and picumeterol (Figure 4.14) by Altria et al. [24]. These workers also noted that as the increase in temperature reduces the viscosity the current and so degree of internal Joule heating will also rise.

Changes in temperature can alter the resolution because of changes to the size of the binding constants and because of

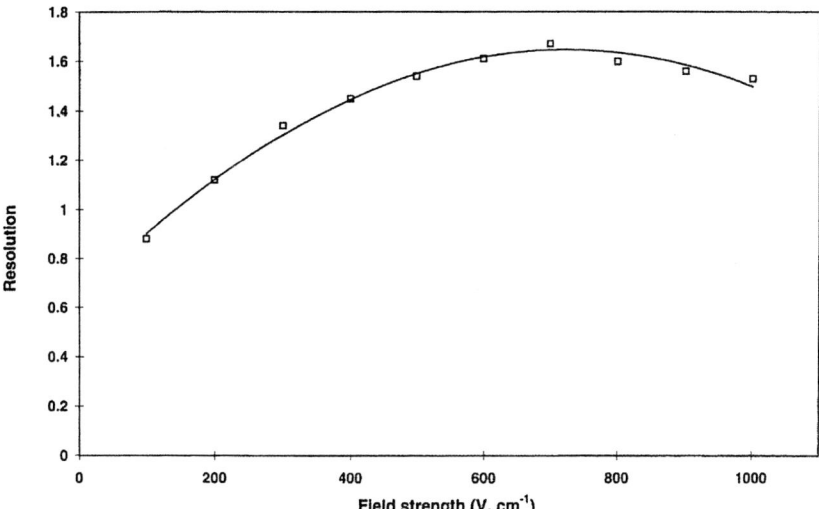

Figure 4.21. The effect of the field strength on the resolution between the enantiomers of Propranolol. Figure adapted from [30] with permission from Elsevier Science.

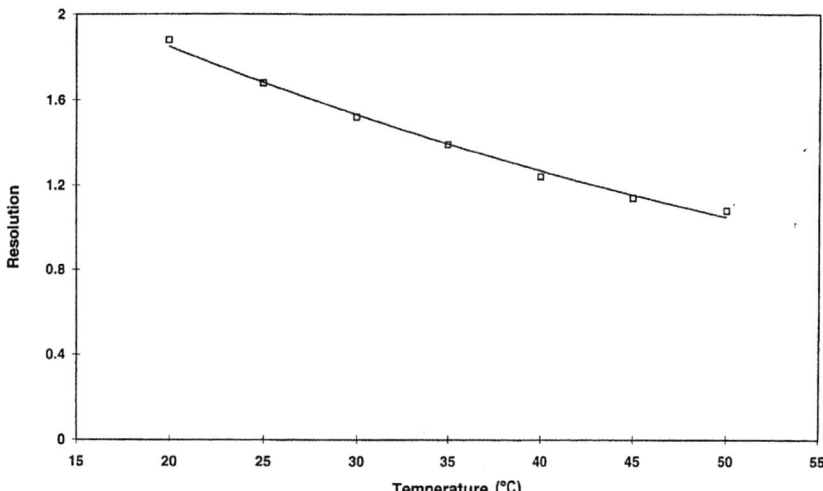

Figure 4.22. The effect of temperature on the resolution between the enantiomers of propranolol. Figure adapted from [30] with permission from Elsevier Science.

changes to the degree of enantioselectivity. The binding constants are often reduced by increases in temperature. If the chiral selector concentration has been optimised for a particular temperature a reduction in the size of the binding constants will mean that the chiral selector concentration is now sub optimal. Increasing the temperature can therefore decrease resolution via the same mechanism as the addition of an organic solvent to the buffer. Changes to the underlying enantioselectivity with temperature are more interesting and arise because of the influence of the temperature on the entropy term in the Gibbs free energy relationship. Enantioselectivity and temperature are discussed in more detail later in the book.

4.6 Optimisation Approaches

From the discussion in the previous section it is clear that there are many experimental factors which determine the degree of enantiomeric resolution achieved. Whilst the different factors influence the resolution to different extents a difficult separation may require all of them to be manipulated to achieve an overall global optimum. The optimisation schemes adopted by most workers can be classified as either univariate or multivariate approaches. In the univariate approach the individual factors are considered separately and one factor, such as the chiral selector concentration, will be optimised before the others are considered. In the mul-

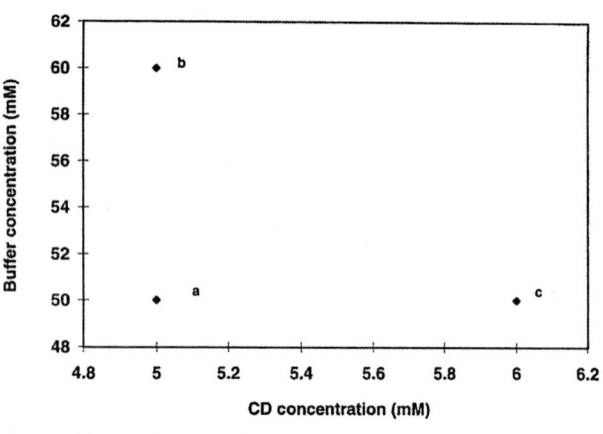 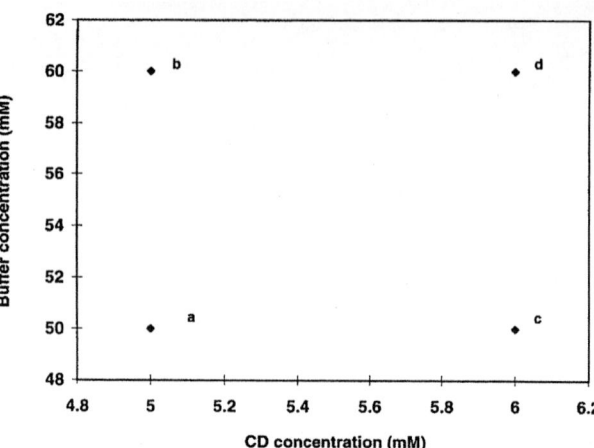

Figure 4.23. Experimental design for two parameters.

tivariate step the various parameters are considered together as part of an overall approach and the experiments are all planned at the beginning and all the data considered together at the end.

4.6.1 Univariate Optimisation

The univariate optimisation procedure is the simplest and the "traditional" approach to method development and is probably still the most widely used. The approach involves the systematic variation of one parameter whilst the others are kept constant. Once an optimum value has been obtained for the first parameter its value is then fixed and the systematic variation applied to the next parameter and so on. Univariate optimisation may be justified by the fact that certain parameters, such as the chiral selector and its concentration, have a more powerful but less predictable influence on enantioselectivity than others such as the temperature.

A common usage of this approach is to choose the chiral selector (for example via the screening procedure detailed earlier), optimise the concentration, and use the other parameters for fine tuning. The univariate approach is implicit in many reports in the literature covering the results of changing experimental conditions. Often the results of changing the cyclodextrin concentration will be given first, followed by the results of changing e.g. the temperature at the optimum cyclodextrin concentration.

Work on the separation of the enantiomers of clenbuterol and picumeterol examined the influence of changing the concentration of β-cyclodextrin, the ionic strength, the pH and buffer composition

and the temperature [24]. Varesio and Veuthey worked on the resolution of the enantiomers of various amphetamines and detailed the results obtained by varying the cyclodextrin type and concentration, the temperature, the voltage, the buffer concentration, the capillary length, and the injection volume [32]. Rickard and Bopp detailed the steps taken to optimise their method for the separation of the enantiomers of a development compound [25]. The chiral selector was chosen after screening a series of cyclodextrins and its concentration was then optimised along with the buffer pH, composition, and concentration. The sample concentration and injection volume were also considered.

4.6.2 Multivariate Optimisation

Whilst the univariate approach is very easy to employ it has the disadvantage that the analytical conditions it produces may not be the optimum ones. The problem arises because several of the experimental parameters used can interact with each other. For example the optimum chiral selector concentration with a 50 mM buffer may not always be the same as that with a 100 mM buffer.

Previous considerations have shown that in many cases there is an optimum chiral selector concentration which varies inversely with the affinity of the analyte for the chiral selector. The affinity is not an absolute value but depends upon environmental factors such as the temperature and the surrounding buffer solution. An increase in buffer concentration for example may increase the affinity. This increase in affinity may mean that the chiral selector concentration is now above the opti-

mum value. Because of the change in affinity, a higher buffer concentration (which might normally be expected to improve the peak shape and so efficiency and resolution) could lead to a reduction in selectivity and so the resolution observed. A higher buffer concentration would also give a higher current and so a greater Joule heating and possibly a higher capillary temperature.

There is often a reluctance amongst analysts to consider the multivariate approach as it contradicts basic instructions given earlier in their education to change only one parameter at a time in an experiment. The objection is based upon the premise that if we change two parameters and see a change in the measured result (e.g. the resolution between the enantiomers) we cannot determine which parameter caused the change. Against this basic objection must be set the knowledge that experimental parameters can interact with one another and so studying them individually in isolation from each other is unlikely to produce the optimum solution.

In order to understand the influence of the parameters we need to design the experiments such that the changes be deconvoluted. To consider a simple case we might want to determine whether the resolution obtained with 5 mM cyclodextrin and 50 mM buffer could be improved by increasing the cyclodextrin concentration to 6 mM or the buffer concentration to 60 mM. Using the univariate approach we would have three experiments a) using 5 mM and 50 mM, b) 5 mM and 60 mM and c) 6 mM and 50 mM (Figure 4.23a). With an experimental design approach we would also have experiment d) using both 6 mM and 60 mM (Figure 4.23b).

The advantage with the experimental design approach is that we measure the change produced by increasing the cyclodextrin concentration at *both* buffer concentrations, and that of increasing the buffer concentration at *both* cyclodextrin concentrations. Experimental design can be thought of in terms of "mapping" parameter space. In this case two parameters were examined at two levels giving rise to four experiments. In general the number of experiments required will be l^n, where l is the number of levels and n the number of parameters. It can be seen that the number of experiments required in this approach increases rapidly with the number of parameters and levels. Because of the number of experiments involved more complex procedures usually involve so called fractional designs in which only a small number of the possible combinations are examined. For a more detailed discussion of experimental design the reader is revered to more specialist texts such as reference [33].

Several workers have used multivariate approaches to look at systems that they had previously examined on a univariate basis. Rogan and co-workers used a Plackett-Burman experimental design to determine the influence of five parameters at three levels on the resolution of clenbuterol enantiomers [34]. The influence of varying the pH, injection time, β-cyclodextrin, buffer, and methanol concentrations upon the resolution, plate count, and migration time were determined. The starting point for the work was the information gained using previous experiments [24]. This previous data was used to set nominal experimental conditions which it was known would produce some resolution. The parameters were examined at high and low levels relative to these initial method conditions with the pH for example being examined at 2.5 and 5.5 in comparison to the method value of 4.0. Fifteen experiments were run and the results used to determine the average results of changing the parameters. The authors conclude that the experimental design gave the same conclusions to those obtained from the univariate optimisation but using a smaller number of experiments.

Bonkerd and collaborators used a Plackett-Burman design to optimise the separation of the enantiomers of fenfluramine [35]. Previously published conditions from the literature were used as a starting point to investigate the affects produced by changing the cyclodextrin and metha-nol concentrations, the pH, voltage and operating temperature. The parameters were studied at two levels using eight experiments. The affect upon the resolution of changing the parameters was determined by subtracting the average result at the low level from the average result at the high level as shown in equation (4.7).

$$E_x = \frac{\sum Y_{x(+1)}}{4} - \frac{\sum Y_{x(-1)}}{4} \qquad (4.7)$$

Where $Y_{x(+1)}$ are the resolution values obtained with parameter x at the high level and $Y_{x(-1)}$ are the resolution values obtained with parameter x at the low level.

With the results obtained from the experimental design the authors were able to improve upon the original method by obtaining a higher resolution in a shorter time.

Other workers have employed more complex procedures such as central composite designs to optimise methods. The central composite design requires many more experiments but can cope with non linear responses and interactions between parameters. Varesio and co-workers used a central composite design to optimise the separation of the enantiomers of amphetamine and four amphetamine derivatives [36]. An established method which had been arrived at by univariate means was used as the basis for the optimisation. The chiral selector and buffer concentrations, the pH, temperature, and voltage were all examined at three levels in twenty eight experiments. In each experiment four responses were determined: the resolution between the amphetamine enantiomers, the sum of the resolutions of all the analyte enantiomers, the analysis time, and the power generated. Multiple regression was used to develop four quadratic equations describing the relationship between the experimental parameters and the responses. The equations developed were used to generate surfaces showing the four responses as a function of both chiral selector concentration and temperature. The final optimised method had the same resolution as the original but with a significant reduction in analysis time. Experimental design was also used to generate response surfaces and optimise the separation of epinephrine enantiomers [37].

4.6.3 Suggested Approaches

The development of an enantioselective method using capillary electrophoresis is not an exact science. In particular it is difficult to predict accurately what the best chiral selector and operating conditions will be. Because of the different enantioselectivities obtained with different chiral selectors and at different concentrations many groups have emphasised the importance of screening a range of selector types and concentrations. The cyclodextrin type and concentration was shown to have a more profound influence on enantioresolution than changes to the buffer ion, temperature or methanol concentration in the buffer [38].

Crommen and co-workers have suggested a method development strategy based upon screening five substituted β-cyclodextrins [39]. The strategy uses a buffer prepared from 100 mM phosphoric acid adjusted to pH 3 using triethanolamine, at a temperature of 15 °C. Different procedures are given in flow charts according to whether the analyte is either basic or neutral/acidic. In the screening experiments the different substituted cyclodextrins are each examined at a single concentration.

With basic analytes the carboxy methyl and sulfobutyl derivatives are screened at a concentration of 5 mM, and the dimethyl, trimethyl, and hydroxypropyl derivatives at a concentration of 15 mM. The cyclodextrin which gives the best resolution in the screening experiments is then examined further at a larger range of concentrations between 1 and 50 mM. If none of the cyclodextrins gives satisfactory resolution but the carboxymethyl derivatives shows some selectivity it is examined again at a pH of 5. If none of these steps are successful, and the addition of an organic modifier does not improve the situation, the authors suggest the use of another type of chiral selector such as a crown ether, antibiotic or protein.

With anionic or neutral analytes the carboxy methyl and sulfobutyl derivatives are screened at concentrations of 10 and 5 mM respectively. If the first step does not produce satisfactory resolution the authors recommend using dual cyclodextrin systems containing 10 mM of dimethyl or trimethyl-β-cyclodextrin in addition to the carboxy methyl and sulfobutyl derivatives. If the fixed concentration dual systems do not work the concentration of the neutral cyclodextrin should be examined in the range 1–50 mM. If variation in the concentrations of neutral cyclodextrins and addition of an organic modifier does not give satisfactory resolution the authors recommend a different class of chiral selector.

The strategy was applied to 30 basic racemic drugs and resolution values between 1.8 and 24 were obtained for 28 of them, with the carboxy methyl derivative being used most often. The application of the strategy to 20 neutral and acidic drugs resulted in success in every case with resolution values of between 1.9 and 31 being obtained. Dual cyclodextrin systems were used on each occasion with the pH 5 buffer being used for eight of the analytes.

A flow chart for method development using substituted α, β, and γ-cyclodextrins was suggested by Roos and co-workers [40]. In initial screening experiments the cyclodextrins are used at fixed concentrations. If the resolution achieved at the fixed concentration exceeds a threshold value the concentration can then be optimised. Different parts of the flow chart are followed according to whether the analytes are charged or electrically neutral. Charged analytes are screened using β-cyclodextrin and the dimethyl and hydroxy propyl derivatives. If the β-cyclodextrins do not produce sufficient resolution then α and γ-cyclodextrin are examined along with their dimethyl and hydroxy propyl derivatives. The α-cyclodextrins are selected for analytes which lack an aromatic ring or contain a long alkyl chain and γ-cyclodextrins for analytes containing condensed aromatic rings. If none of the neutral cyclodextrins give enough resolution with the charged analytes then charged cyclodextrins are employed. Neutral analytes are screened with either carboxy methyl or amino β-cyclodextrins.

Liu and Nussbaum developed a screening strategy for basic analytes based upon dimethyl-β-cyclodextrin, sulphonated β-cyclodextrin, and hydroxypropyl α, β, and γ-cyclodextrins [41]. They examined twenty commercially available compounds and eighteen pharmaceutical development compounds using the cyclodextrins at two different concentrations. The buffer system used 25 mM phosphoric acid adjusted to pH 2.5 using either tetrabutyl ammonium hydroxide or triethylamine. In the initial work partial or complete resolution was obtained for all of the enantiomers and 50% gave baseline resolution. The initial work showed that whilst different levels of success were achieved,

none of the cyclodextrins were successful for all analytes and that they often complimented each other. Based upon the initial studies the authors suggest screening with five buffers containing either 15 mM dimethyl-β-cyclodextrin, 30 mM hydroxypropyl α, β, and γ-cyclodextrins, and 32 mM sulphonated β-cyclodextrin. The buffer with the sulphonated β-cyclodextrin is used in a narrower capillary and with a lower voltage because of the high currents that would be generated otherwise. The screening strategy was applied to a further ten development compounds and eight were partially or fully resolved.

Whilst there are a number of procedures and guidelines which can be adopted to maximise the chances of producing a successful method, it is important to remember that the recommendations given are subjective and that other approaches may be at least as valid. The approach given below is that suggested by the author based upon his own work and the results published in the literature.

As it is not currently possible to predict the best chiral selector for any analyte it is recommended that rapid screening techniques are employed so that several can be evaluated quickly. Chiral selectors which show promise can then be examined in greater depth. The other experimental parameters should be chosen to maximise separation efficiency such that any enantioselectivity has the best chance of producing detectable resolution.

Ideally the sample employed for method development will be a racemate or alternatively a sample which has a high level of the minor enantiomer. A dilute sample and small injection volume can then be employed so that the degree of peak tailing will be small and the chances of observing enantioselectivity are maximised.

Rapid screening can be carried out using short narrow capillaries, moderate buffer strengths, and high electrical fields. A capillary which has been coated to limit the electroosmotic flow should be employed. Alternatively a fused silica capillary can be used with low pH buffers. In the first instance a pH should be chosen that ensures that the analyte is fully charged. Analytes which are neutral or do not carry a significant charge between pH 2 and 12 will require a charged chiral selector. Each chiral selector chosen should then be examined in scouting experiments at a wide range of concentrations, for example 0, 0.5, 5 and 50 mM. Cyclodextrins which carry a charge opposite to that on

the analyte may be especially beneficial and should be considered. Monitoring the mobility change at a wide range of concentrations gives an idea of the strength of analyte-selector interaction. This author recommends examining the potential of cyclodextrins rather than other chiral selectors in the first instance. The various merits and demerits of cyclodextrins and other chiral selectors are discussed in some detail in the following chapters.

The chiral selectors which show most promise at this stage should then be picked for further optimisation work. A much more limited range of chiral selector concentrations can then be examined based upon the results of the scouting experiments. If for example there is no evidence of any enantioselectivity at 0.5, and 5 mM but some at 50 mM, we might want to try the additional concentrations of 30, 40, 60, and 70 mM. Monitoring changes in electrophoretic mobility has the benefit of indicating the degree of binding. These mobility changes can be useful as an indicator that whilst binding is occurring it is not enantioselective.

When a chiral selector concentration has been established which is close to the optimum value, changes to the other buffer parameters such as concentration, type of buffer ion and pH can be undertaken. At this stage it may be appropriate to use a sample which is more representative of that which will be analysed routinely e.g. one that is nearly enantiomerically pure. It may also be appropriate to switch to a longer and wider capillary which will maximise the detectability of the minor enantiomer.

Experimental designs are very useful at the fine tuning stage in method development as they give information on how the various parameters interact and so how a global optimum can be achieved. Experimental designs can also give useful information on the robustness of the final method by giving information on the sensitivity of the method to small changes in the various parameters. The measurement of several responses in the experimental design can also help in final decisions about the method parameters e.g. the cost in the run time of a small improvement in resolution.

4.7 Validation

The factors used to validate methods of enantiomer separation by CE are similar to those used in HPLC. The factors determined include: linearity of response, accuracy, precision, robustness, specificity, and determination of the limits of detection (LOD) and quantification (LOQ). Two important points to remember are that the injection volume reproducibility is usually inferior to that seen in HPLC and that peak areas depend on migration times. Because of the variability in injection volume an internal standard is recommended for those methods not based upon area % procedures. The precision in the estimation of the enantiomeric purity of l-epinephrine was shown to be improved by the use of l-pseodoephedrine as an internal standard [42]. With CE the slower moving analytes spend longer in the detection region and so give larger peak areas than faster analytes. In order to compensate the peak areas are usually divided by the analysis times to give corrected areas. An additional complication is that as interaction with a cyclodextrin or other chiral selector can alter the molar extinction coefficient, there is no guarantee that the two enantiomers will give exactly equal responses. Any errors due to different responses are, however, likely to be small in relation to the uncertainties in estimating low levels of the minor enantiomer.

Validation of a method for the enantiomeric purity testing of fluparoxan (Figure 4.24) showed good linearity when each of the pure enantiomers were spiked with 1 – 8% of the opposite enantiomer. The limits of detection and quantification were 0.3% and 1.0% respectively with good precision being obtained at the 1% level [43].

The ability to transfer methods from one site to another and between equipment from different manufacturers was demonstrated in an inter company cross validation exercise reported by Altria and others [44]. Seven pharmaceutical companies at seven locations used equipment from three instrument companies to analyse the same sample of clenbuterol by the same method. All of the companies obtained similar enantioselectivities and similar data for linearity of response, migration time precision, and peak area precision.

A comprehensive validation study on a method to determine the enantiomeric purity of the local anaesthetic (S)-ropiva-

caine has been reported by Sänger-van de Griend and co-workers [45]. The study covered specificity, limits of detection and quantification, linearity, accuracy, repeatability, intermediate precision, robustness, and a comparison with the results from a validated HPLC method.

The levels of the minor enantiomer, (R)-ropivacaine, were determined by an area % method using corrected peak areas as shown in equation (4.8).

$$\% R = \frac{100 \cdot A_R/t_R}{A_S/t_S + A_R/t_R} \qquad (4.8)$$

Where A_R and A_S are the peak areas and t_R and t_S the migration times for the R and S enantiomers.

The specificity was demonstrated by showing that homologs of the ropivacaine enantiomers are resolved by the method. Measurement of the limit of detection was limited by the enantiomeric purity of the standard but was below 0.08%. At the 0.08% level the RSD of six injections was 8%. The linearity was determined in the range 1 – 3% of the R enantiomer and gave an r^2 value of 0.9981 with a zero intercept. The accuracy was determined by showing a linear relationship between the amount of the R enantiomer spiked into the formulation and that determined experimentally. The repeatability was determined at five levels between 0.1% and 2.6% by making six determinations at each level and the RSDs shown to vary between 6.6% and 0.6%. The intermediate precision was determined by the same analyst using the same instrument on three different days. The RSDs were similar to those produced by the repeatability work. The method robustness was determined by using a full factorial design at two levels to determine the influence of changing the pH, buffer concentration, cyclodextrin, and temperature. Satisfactory resolution was obtained even with the combination of parameters which gave the lowest resolution. The CE method was compared with an HPLC method by comparing the results obtained from twenty six solutions generated by stability studies. A paired t-test showed that the two methods did not give significantly different results.

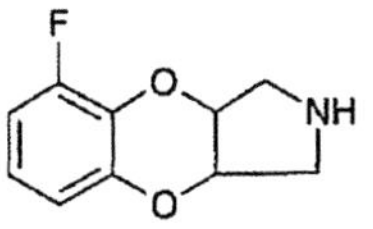

4.8 Computer Simulations

As it is possible to model the electrophoretic mobility of the individual enantiomers based upon their binding constants, the chiral selector concentration, their pK_a

Fluparoxan

Figure 4.24. The structure of fluparoxan.

values, and the buffer pH, the enantioselectivity can be simulated using computer software. Reijenga and co-workers have detailed such a program for a personal computer [46]. The program allows the input of the equilibrium constants for both enantiomers in their charged and neutral forms. Based upon the input experimental parameters the program simulates an electropherogram and allows the user to experiment with different values. Computer simulation programs which simulate the separation of enantiomers in CE are a valuable training tool.

4.9 Conclusion

The development of successful enantiomer separation methods in CE is a process which whilst partly dependant upon experience and luck, is greatly assisted by a rigorous approach. In particular an understanding of the underlying physical processes and interactions gives helpful guidance. The use of a systematic approach to screening chiral selectors and optimising experimental conditions is also beneficial.

References

[1] Mikkers, F.E.P.; Everaerts, F.M.; Verheggen, Th. P.E.M. Concentration distributions in free zone electrophoresis, *J. Chromatogr.* **1979**, *169*, 1 – 10.

[2] Terabe, S.; Yashima, T.; Tanaka, N.; Araki, M. Separation of oxygen isotope benzoic acids by capillary zone electrophoresis based on isotope effects on the dissociation of the carboxyl group, *Anal. Chem.* **1988**, *60*, 1673 – 1677.

[3] Hutterer, K.M.; Jorgenson, J.W. Ultrahigh-Voltage Capillary Zone Electrophoresis, *Anal. Chem.* **1999**, *71*, 1293 – 1297.

[4] Giddings, J.C. Generation of Variance, "Theoretical plates", Resolution, and Peak Capacity in Electrophoresis and Sedimentation, *Sep. Sci.* **1969**, *4*, 181 – 189.

[5] Kenndler, E.; Schwer, C. Nondependence of Diffusion-Controlled Peak Dispersion on Diffusion Coefficient and Ionic Mobility in Capillary Zone Electrophoresis without Electroosmotic Flow, *Anal. Chem.* **1991**, *63*, 2499 – 2502.

[6] Sänger-van de Griend, C.E.; Gröningsson, K.; Westerlund, D. Chiral Separation of

Local Anaesthetics with Capillary Electrophoresis. Evaluation of the Inclusion Complex of the Enantiomers with Heptakis(2,6-di-O-methyl)-β-cyclodextrin, *Chromatographia* **1996**, *42*, 263–268.

[7] Rogan, M.M.; Altria, K.D.; Goodall, D.M. Enantiomeric separation of Salbutamol and related impurities using capillary electrophoresis, *Electrophoresis* **1994**, *15*, 808–817.

[8] Wren, S.A.C. Theory of chiral separation in capillary electrophoresis, *J. Chromatogr.* **1993**, *636*, 57–62.

[9] Armstrong, D.W.; Rundlett, K.; Reid III, G.L. Use of a Macrocyclic Antibiotic, Rifamycin B, and Indirect Detection for the Resolution of Racemic Amino Alcohols by CE, *Anal. Chem.* **1994**, *66*, 1690–1695.

[10] Fillet, M.; Bechet, I.; Chiap, P.; Hubert, Ph.; Crommen, J. Enantiomeric determination of propranolol by cyclodextrin-modified capillary electrophoresis, *J. Chromatogr. A* **1995**, *717*, 203–209.

[11] Penn, S.G.; Bergström, E.T.; Goodall, D.M.; Loran, J.S. Capillary Electrophoresis with Chiral Selectors: Optimization of Separation and Determination of Thermodynamic Parameters for Binding of Tioconazole Enantiomers to Cyclodextrins, *Anal. Chem.* **1994**, *66*, 2866–2873.

[12] Chiari, M.; Desperati, V.; Manera, E.; Longhi, R. Combinatorial Synthesis of Highly Selective Cyclohexapeptides for Separation of Amino Acid Enantiomers by Capillary Electrophoresis, *Anal. Chem.* **1998**, *70*, 4967–4973.

[13] Guttman, A.; Brunet, S.; Cooke, N. Capillary Electrophoresis Separation of Enantiomers Using Cyclodextrin Array Chiral Analysis, *LC-GC Intl.* February **1996**, 88–100.

[14] Guttman, A.; Jurado, C.; Brunet, S.; Cooke, N. Rapid Chiral Separations Methods Development by Cyclodextrin-Mediated Capillary Electrophoresis for Acidic and Basic Compounds, *Chirality* **1995**, *7*, 409–419.

[15] Guttman, A. Novel separation scheme for capillary electrophoresis of enantiomers, *Electrophoresis* **1995**, *16*, 1900–1905.

[16] Aumatell, A.; Guttman, A. Ultra-fast chiral separation of basic drugs by capillary electrophoresis, *J. Chromatogr. A.* **1995**, *717*, 229–234.

[17] Rawjee, Y.Y.; Staerk, D.U.; Vigh, G. Capillary electrophoretic separations with cyclodextrin additives. I. Acids: chiral selectivity as a function of pH and the concentration of β-cyclodextrin for fenoprofen and ibuprofen, *J. Chromatogr.* **1993**, *635*, 291–306.

[18] Baumy, Ph.; Morin, Ph.; Dreux, M.; Viaud, M.C.; Boye, S.; Guillaumet, G. Determination of β-cyclodextrin inclusion complex constants for 3,4-dihydro-2-*H*-1-benzopyran enantiomers by capillary electrophoresis, *J. Chromatogr. A* **1995**, *707*, 311–326.

[19] Perrin, D.D.; Dempsey, B. Buffers for pH and Metal Ion Control, Chapman and Hall, **1974**, London.

[20] Bello, M.S. Electrolytic modification of a buffer during a capillary electrophoresis run, *J. Chromatogr. A* **1996**, *744*, 81–91.

[21] Hows, M.E.P.; Perrett, D. Effects of Buffer Depletion in Capillary Electrophoresis:

Development of a Continuous flow Cathode, *Chromatographia* **1998**, *48*, 355–359.

[22] Macka, M.; Andersson, P.; Haddad, P.R. Changes in Electrolyte pH Due to Electrolysis during Capillary Zone Electrophoresis, *Anal. Chem.* **1998**, *70*, 743–749.

[23] Mikkers, F.E.P.; Everaerts, F.M.; Verheggen, Th.P.E.M. Concentration distributions in free zone electrophoresis, *J. Chromatogr.* **1979**, *169*, 1–10.

[24] Altria, K.D.; Goodall, D.M.; Rogan, M.M. Chiral Separation of β-Amino Alcohols by Capillary Electrophoresis Using Cyclodextrins as Buffer Additives. I. Effect of Varying Operating Parameters, *Chromatographia* **1992**, *34*, 19–24.

[25] Rickard, E.C.; Bopp, R.J. Optimization of a capillary electrophoresis method to determine the chiral purity of a drug, *J. Chromatogr. A* **1994**, *680*, 609–621.

[26] Sustácek, V.; Foret, F.; Bocek, P. Selection of the background electrolyte composition with respect to electromigration dispersion and detection of weakly absorbing substances in capillary zone electrophoresis, *J. Chromatogr.* **1991**, *545*, 239–248.

[27] Stålberg, O.; Hedeland, M.; Pettersson, C.; Westerlund, D. The Effect of Conductivity Tuning in Chiral Separations by CE: Using Hydroxypropyl-β-Cyclodextrin in Combination with Tetraalkylammonium Ions, *Chromatographia* **1998**, *48*, 415–421.

[28] Williams, R.L.; Vigh, G. Maximization of separation efficiency in capillary electrophoretic separations by means of mobility-matching background electrolytes, *J. Chromatogr. A* **1996**, *730*, 273–278.

[29] Williams, R.L.; Vigh, G. Polyethylene glycol monomethyl ether sulfate-based background electrolytes in capillary electrophoresis, *J. Chromatogr. A* **1996**, *744*, 75–80.

[30] Guttman, A.; Cooke, N. Practical aspects of chiral separations of pharmaceuticals by capillary electrophoresis I. Separation optimization, *J. Chromatogr. A* **1994**, *680*, 157–162.

[31] Nielen, M.W.F. Chiral Separation of Basic Drugs Using Cyclodextrin-Modified Capillary Zone Electrophoresis, *Anal. Chem.* **1993**, *65*, 885–893.

[32] Varesio, E.; Veuthey, J.-L. Chiral separation of amphetamines by high-performance capillary electrophoresis, *J. Chromatogr. A* **1995**, *717*, 219–228.

[33] Montgomery, D.C. Design and analysis of experiments 4th edition, J. Wiley & Sons, New York, **1997**.

[34] Rogan, M.M.; Altria, K.D.; Goodall, D.M.; Plackett-Burman Experimental Design in Chiral Analysis using Capillary Electrophoresis, *Chromatographia* **1994**, *38*, 723–729.

[35] Boonkerd, S.; Detaevernier, M.R.; Vander Heyden, Y.; Vindevogel, J.; Michotte, Y. Determination of the enantiomeric purity of dexfenfluramine by capillary electrophoresis: use of a Plackett-Burman design for the optimization of the separation, *J. Chromatogr. A* **1996**, *736*, 281–289.

[36] Varesio, E.; Gauvrit, J.-Y.; Longeray, R.; Lantéri, P.; Veuthey, J.-L. Central composite design in the chiral analysis of amphetamines by capillary electrophoresis, *Electrophoresis* **1997**, *18*, 931–937.

[37] Fanali, S.; Furlanetto, S.; Aturki, Z.; Pinzauti, S. Experimental Design Methodologies to Optimize the CE Separation of Epinephrine Enantiomers, *Chromatographia* **1998**, *48*, 395–401.

[38] Bechet, I.; Paques, P.; Fillet, M.; Hubert, P.; Crommen, J. Chiral separation of basic drugs by capillary zone electrophoresis with cyclodextrins additives, *Electrophoresis* **1994**, *15*, 818–823.

[39] Fillet, M.; Hubert, P.; Crommen, J. Method development strategies for the enantioseparation of drugs by capillary electrophoresis using cyclodextrins as chiral additives, *Electrophoresis* **1998**, *19*, 2834–2840.

[40] Roos, N.; Ganzler, K.; Szemán, J.; Fanali, S. Systematic approach to cost- and time-effective method development with a starter kit for chiral separations by capillary electrophoresis, *J. Chromatogr. A* **1997**, *782*, 257–269.

[41] Liu, L.; Nussbaum, M.A. Systematic screening approach for chiral separations of basic compounds by capillary electrophoresis with modified cyclodextrins, *J. Pharm. Biomed. Anal.* **1999**, *19*, 679–694.

[42] Peterson, T.E.; Trowbridge, D. Quantitation of *l*-epinephrine and determination of the *d*-*ll*-epinephrine enantiomer ratio in a pharmaceutical formulation by capillary electrophoresis, *J. Chromatogr.* **1992**, *603*, 298–301.

[43] Altria, K.D.; Walsh, A.R.; Smith, N.W. Validation of a capillary electrophoresis method for the enantiomeric purity testing of fluparoxan, *J. Chromatogr.* **1993**, *645*, 193–196.

[44] Altria, K.D.; Harden, R.C.; Hart, M.; Hevizi, J.; Hailey, P.A.; Makwana, J.V.; Portsmouth, M.J. Inter-company cross-validation exercise on capillary electrophoresis, *J. Chromatogr.* **1993**, *641*, 147–153.

[45] Sänger-van de Griend, C.E.; Wahlstrom, H.; Gröningsson, H.; Widahl-Näsman, M. A chiral capillary electrophoresis method for ropivacaine hydrochloride in pharmaceutical formulations: validation and comparison with chiral liquid chromatography, *J. Pharm. Biomed. Anal.* **1997**, *15*, 1051–1061.

[46] Reijenga, J.C.; Ingelse, B.A.; Everaerts, F.M. Training software for chiral separations in capillary electrophoresis, *J. Chromatogr. A* **1997**, *772*, 195–202.

The Use of Cyclodextrins as Chiral Selectors

5.1 Introduction

Cyclodextrins are the most popular of the many chiral selectors used in CE because they have many of the desirable features of the ideal chiral selector. Cyclodextrins are used in about two thirds of the literature applications of CE for the separation of enantiomers [1]. Cyclodextrins show good enantioselectivity for a wide range of analytes, are transparent to UV light down to low wavelengths, and have good water solubility. Cyclodextrins are available in a range of sizes and chemistries, usually give fast kinetics for the formation and breakdown of complexes with enantiomers, and are relatively cheap. Cyclodextrins were amongst the first chiral selectors employed in CE and their successful application has followed their use as chiral stationary phases in GC, TLC, and HPLC, and as mobile phase additives in TLC and HPLC. Most early workers used the parent α, β, and γ-cyclodextrins but most interest has now shifted to the substituted cyclodextrin derivatives, particularly those of β-cyclodextrin. Charged cyclodextrins are rapidly growing in popularity with anionic derivatives such as various sulphonated β-cyclodextrin being the most widely used.

The use of cyclodextrins as chiral selectors in CE has grown rapidly over the last decade with an almost exponential increase in the number of publications [2]. The change in the number of publications per year since 1990 is shown in Figure 5.1.

This growth in the number of publications reflects both the increase in the interest in CE as an analytical technique generally and the recognition of the benefits of using CE with cyclodextrins to perform enantioselective separations.

Cyclodextrins are cyclic oligosaccharides composed of D-glucose units connected through the 1 and 4 positions by α linkages. The structure of the closed ring form of glucose is shown below in Figure 5.2.

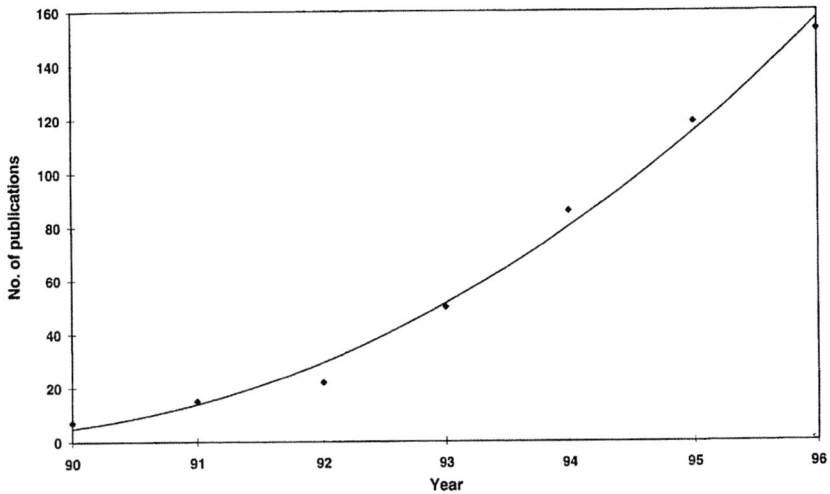

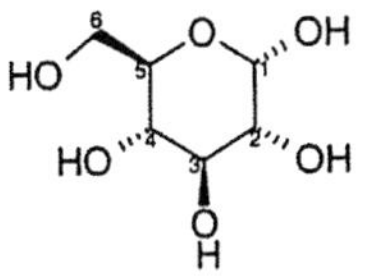

Figure 5.1. The increase in the number of publications citing the use of cyclodextrins in CE.

The α-cyclodextrin contains six glucose units, β-cyclodextrin seven glucose units, and γ-cyclodextrin eight glucose units. Cyclodextrins are usually described and depicted as either doughnut shaped or slightly conical cylindrical molecules. Cyclodextrins have a relatively hydrophobic cavity in the middle and a relatively hydrophilic surface on the outside. These differences in internal and external hydrophobicities mean that cyclodextrins have the ability to accept smaller host molecules into their cavity.

5.2 Structure and Properties

The solid state structures of the α, β, and γ-cyclodextrins have been determined by numerous spectroscopic techniques including X-ray diffraction, solid state NMR, Infra red, and Raman spectroscopy [3]. Cyclodextrin structures in solution have been studied by both proton and ^{13}C NMR. The chemical shifts and coupling constants of the cyclodextrin and guest analyte may change upon complexation, and these changes can be used to infer information about the orientation

Figure 5.2. The closed ring form of D-glucose.

of the guest molecule within the cyclodextrin cavity. Cyclodextrin purity may be monitored by chromatographic techniques such as HPLC and TLC [3]. The structure of β-cyclodextrin is shown below in Figure 5.3.

The glucose units in β-cyclodextrin are joined through the 1 and 4 positions and this leads to the primary hydroxyl groups being located on one side of the β-cyclodextrin and the secondary hydroxyl groups on the other. The α and γ-cyclodextrins have similar structures and differ only in the number of glucose units. The cyclodextrins molecules have cavity depths of 0.78 nm and cavity widths of 0.57, 0.78, and 0.95 nm respectively [2]. The cyclodextrins differ not only in their physical dimensions but also in some of their physico chemical properties. Some of the properties of the cyclodextrins are shown in Table 5.I below [3].

0009-5893/00/02   59-19 $ 03.00/0

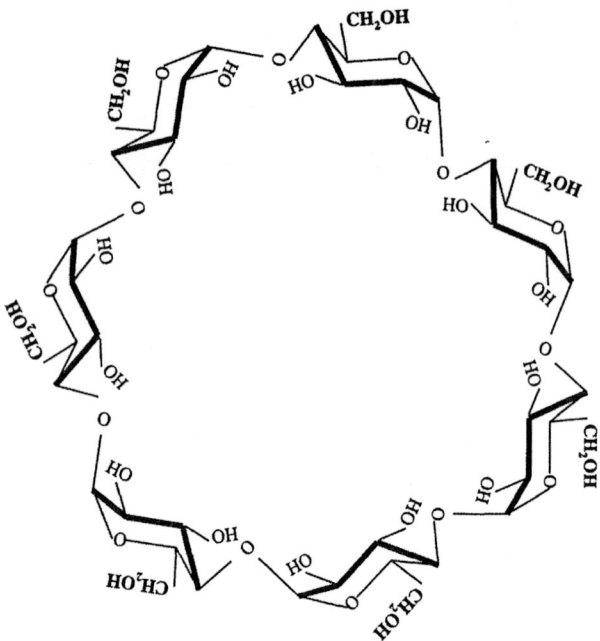

Figure 5.3. The structure of β-cyclodextrin.

Table 5.I. Some properties of the α, β, and γ-cyclodextrins.

Property	α	β	γ
Molecular weight	972	1135	1297
Solubility in water (mg/mL)	145	18.5	232
pK_a		12.3	12.2	12.1
$[\alpha]_d$		150	163	177

5.3 Cyclodextrin Production

Cyclodextrins are produced by the action of a bacterial enzyme known as cyclodextrin glucosyl transferase (CGTase) on starch solutions [2, 3]. The enzyme is added to a partially hydrolysed starch solution and all three common cyclodextrins can be produced depending upon the reaction conditions. The degree of starch hydrolysis is important as if it is too high the main product is glucose and other low molecular weight oligosaccharides. The proportions of the different cyclodextrins produced can vary with the source of the CGTase and the time of fermentation. The proportions of the different cyclodextrins may also be altered by the use of various reagents to isolate the component of interest. Toluene leads to the formation of a toluene/β-cyclodextrin complex which pushes the equilibrium towards the generation of more β-cyclodextrin. The use of 1-decanol favours the production of α-cyclodextrin, and with cyclohexadecanol the main species formed is γ-cyclodextrin [3].

The ability of cyclodextrins to include other molecules as guests has lead to much interest in the use of cyclodextrins as a means to modify the behaviour of the guest molecules. Cyclodextrins have been used to increase the aqueous solubility of hydrophobic drugs as a way of improving the absorption rates. The aqueous solubility of the diuretic Spironolactone was increased eight and seventeen fold by the use of β-cyclodextrin and γ-cyclodextrin respectively [9]. The use of γ-cyclodextrin was shown to increase the serum levels of diazepam following oral administration to rabbits [10]. Cyclodextrins are also used in a wide range of other applications such as the processing of foods and the manufacture of cosmetics and toiletries [2]. The widespread interest in the use of cyclodextrins as pharmaceutical excipients, solubilisers, additives, catalysts etc. has been the driving force behind the major improvements in the fermentation and purification processes. Because of this demand for industrial scale production many cyclo-

One of the most interesting features of the native α, β, and γ-cyclodextrins is the difference in their water solubilities. The aqueous solubility of β-cyclodextrin at ambient temperature is one order of magnitude less than that of the α, and γ-cyclodextrins. One possible explanation for the large solubility differences is the difference in the nature of the hydrogen bonds formed between adjacent glucose units in the cyclodextrin ring. The hydroxyl group on carbon atom 2 of one of the glucose units can form a hydrogen bond with the hydroxyl group on carbon atom 3 in the adjacent glucose unit. With β-cyclodextrin all of these opportunities for intramolecular hydrogen bonds are believed to be taken, leading to a relatively rigid structure. With the other cyclodextrins the geometry differences mean that there are fewer intramolecular hydrogen bonds and so presumably more interactions with the surrounding water molecules [2, 4]. The connection between intramolecular hydrogen bonds and the limited solubility of β-cyclodextrin is supported by the greater solubility of many β-cyclodextrin derivatives. The increase in solubility is even seen with the ether derivatives such as methyl and dimethyl-β-cyclodextrin. The aqueous solubilities of the hydroxypropyl, acetyl, and different methyl derivatives of β-cyclodextrin are typically one order of magnitude higher than that of the parent β-cyclodextrin [5]. For most groups of compounds an ether derivative could nor-

mally be expected to have a lower solubility in aqueous media than that of the parent alcohol. The solubility of β-cyclodextrin itself can be improved by the use of various additives such as organic solvents or high concentrations of urea. The solubility of β-cyclodextrin in solutions containing variable proportions of water and organic solvents such as methanol, ethanol, propanol, acetonitrile, THF and DMSO has been determined [6, 7]. The change in solubility of β-cyclodextrin was found to vary greatly with the type and concentration of the organic solvent added. With most of the solvents the solubility initially increased with increasing concentration of organic solvent but then declined at higher concentration, usually to below the value obtained in pure water. The exception to this pattern of an increase in solubility at low concentrations of organic solvent was methanol. The solubility of β-cyclodextrin in methanol/water mixtures was always lower than that found in pure water. With the other alcohols, acetonitrile and THF the use of concentrations in the range of up to about 12% gave a modest increase in the solubility of β-cyclodextrin, and higher solvent concentrations gave a decline in solubility. The use of a 4 M urea solution increases the solubility of β-cyclodextrin by a factor of five and an 8 M urea solution by a factor of thirteen [8].

dextrins have changed in status from that of fine chemicals to that of bulk commodities. The cost of β-cyclodextrin for example has changed from about $ 2000 per kg in 1970 to a few dollars per kg in 1995 [2].

The derivatised cyclodextrins are mostly synthesised by simple chemical modification of the hydroxyl groups on the parent cyclodextrin. Many of the derivatisation reactions used are non specific and because there are two secondary hydroxyl groups and one primary hydroxyl group per glucose unit there are many potential sites for substitution (18, 21 and 24 for α, β, and γ-cyclodextrins respectively). Because of the lack of selectivity most of the derivatised cyclodextrins contain many different components varying in both the degree and position of substitution. As the different components can give rise to different degrees of enantioselectivity the batch to batch reproducibility of derivatised cyclodextrins can be an important issue and this is covered later in the chapter.

5.4 Complexation Mechanisms

5.4.1 Equilibrium Constants

The enantioselectivity expressed by cyclodextrins is usually thought to arise via either or both of two mechanisms: i) the differential inclusion of the two enantiomers into the cyclodextrin cavity, and ii) the differential behaviour of the two inclusion complexes formed between the cyclodextrin and the two enantiomers. There is a great deal of evidence for the formation of inclusion complexes with many analytes in the aqueous buffer systems used in CE. Whilst inclusion has been shown to occur with many analytes there may also be situations in which non inclusion enantioselective interactions are also important.

The size of the equilibrium constants for the formation of analyte-cyclodextrin complexes depends both upon the size of the analyte and the size of the cyclodextrin. An important factor seems to be the need for a good match between the size of the analyte and the size of the cavity. In simple terms if the analyte is too large it cannot fit easily into the cyclodextrin cavity and if it is too small it will not be bound tightly enough. The size matching considerations are most relevant for one to one complexes, i.e. those where a single ana-

Table 5.II. Formation constants.

Analyte	α	β	γ
1-Butanol	88	17	–
Iodine	8,400	6	–
Phenol	15,200	2,400	–
Cyclohexanol	64	480	–
Benzene	32	170	9
2-Naphthalenesulphonate	363	234,400	38
2,7-Naphthalenedisulphonate	10	275	380

lyte molecule is included into a single cyclodextrin molecule. Complexation stoichiometries other than one to one are also possible: γ-cyclodextrins may be able accommodate two guest molecules at the same time, and smaller cyclodextrins may be able to complex simultaneously with two or more hydrophobic sites on the same analyte. An example of some small peptides complexing with more than one cyclodextrin molecules is discussed later in this chapter.

Table 5.II shows the equilibrium constants for the formation of 1:1 complexes between α, β, and γ-cyclodextrins and some achiral analytes [11].

From the data shown in Table 5.II it can be seen that in general small analytes complex the most strongly with α-cyclodextrin, medium sized analytes most strongly with β-cyclodextrin, and the larger analytes most strongly with γ-cyclodextrin.

Many of the common pharmaceutical agents have sizes such that they interact more strongly with β-cyclodextrin and γ-cyclodextrin rather than α-cyclodextrin. Many of the benzodiazepines for example show stronger affinity for β-cyclodextrin [10], and many of the steroids show stronger affinity for the larger γ-cyclodextrin [12].

It is sometimes assumed that the mechanism for the inclusion of an analyte into the cyclodextrin is driven by hydrophobic interactions between the analyte, or a part of the analyte, and the cyclodextrin cavity. Whilst hydrophobic interactions may be significant in some cases other interactions such as steric or polar forces may be more important in others. With the case of benzodiazepines and γ-cyclodextrin hydrophobicity appears to be a major driving force. A linear relationship was found between the log of the stability constant and the log of the octanol-water partition coefficient for 13 benzodiazepines [10]. In other cases the relationship between stability complexes and log P is more complex. The relationship be-

tween the octanol-water partition coefficient and the stability complex equilibrium constants were examined for eighteen steroids and α, β, and γ-cyclodextrin [12]. For α and γ-cyclodextrin there was no correlation at all between the stability constants and the octanol-water partition coefficient and for β-cyclodextrin an r^2 value of only 0.42 was seen.

5.4.2 Evidence for Inclusion

The evidence for the formation of inclusion complexes between analytes and cyclodextrins in both the solid phase and in solution comes from a number of sources. X-ray crystallography and X-ray powder diffraction have both been used to study complexes in the solid state [3]. The structure of the 1:1 complex formed between 2,6-di-o-methyl-β-cyclodextrin and adamantol was determined by X-ray crystallography [13].

With the solution studies some of the most compelling evidence for inclusion comes from changes to the ^{1}H-NMR shifts for the different glucose proton signals within the cyclodextrin rings. The ^{1}H-NMR shifts were measured for α, β, and γ-cyclodextrin in both the presence and absence of hydrocortisone [12]. With β and γ-cyclodextrin, which both have large equilibrium constants, there were significant shifts to the H_3 and H_5 proton signals and smaller shifts to the H_6 signals (see Figure 5.2). The signals for the H_1, H_2, and H_4 protons which lie on the outside of the cyclodextrin rings were shifted to smaller extents. In the case of α-cyclodextrin (which forms a much weaker complex) the H_3 signal is shifted the most with the changes to the other signals all being similar.

Similar changes to the shifts of the ^{1}H-NMR signals in β-cyclodextrin were seen when the β-blocker propranolol was also present in solution [14]. The shift in the signals rather than the appearance of new signals indicated that an average environment was being observed with very rapid formation and break up of the complex. Again the signals due to the H_3, H_5, and H_6 protons on the inside of the ring were shifted more than those of H_1, H_2, and H_4 protons on the outside. The size of the shifts depended upon the molar ratio of β-cyclodextrin to propranolol, with the relationship with concentration indicating the formation of a 1:1 complex. Changes to the proton signals on the naphthalene ring of propranolol also indicated that inclusion had occurred. Further information on which of the cyclodextrin protons were physically close to the naphthalene protons was obtained by the use of Nuclear Overhauser experiments. Some of the ^1H signals due to propranolol were seen to be split into two upon inclusion into the cyclodextrin cavity. These split signals were shown to be due to the two enantiomeric forms of propranolol by repeating the experiment using a sample of pure (–) propranolol, and other samples containing different ratios of the two enantiomers. These differences in proton shifts for the two propranolol enantiomers indicated that the two enantiomers have slightly different alignments in the β-cyclodextrin cavity. The differences in shifts for the two enantiomers provide the basis for methods for determining the optical purity of a sample by ^1H-NMR. Similar differences between the shifts for the two enantiomers were also seen following the addition of γ-cyclodextrin but not for α-cyclodextrin.

The changes to the ^1H shifts of β-cyclodextrin upon complexation with the enantiomers of the dinitrophenyl derivatives of valine, leucine, and methionine were studied [15]. The proton shifts due to the H_3 and H_5 protons changed significantly as a function of the molar ratio of analyte to cyclodextrin whereas the proton shifts due to the other cyclodextrin protons changed only slightly. The ^1H shifts of the glucose H_3 proton were measured for different ratios of the cyclodextrin and the different analytes. The rates of change of the shifts and the limiting values were different for the different amino acids and their enantiomers. These differences suggested both differences in the equilibrium constants for the formation of the inclusion complexes and differences in the orientation of the enantiomers and amino acids in the cyclodextrin cavity.

Both ^1H and ^{13}C-NMR spectroscopies were used to study the inclusion into β-cyclodextrin and carboxymethyl β-cyclodextrin of the antihistamine dimethindene [16]. Some of the dimethindene ^1H signal shifts changed as a function of the ratios of dimethindene to the cyclodextrins. In addition some of the ^1H and ^{13}C-NMR shifts were split into the signals due to the two enantiomers upon complexation with the cyclodextrins. Comparison of the results obtained with the two cyclodextrins showed that different dimethindene signals were split upon complexation suggesting a different alignment in the cyclodextrin cavities.

5.4.3 The Role of the Organic Solvent

The degree of inclusion of the analyte into the cyclodextrin cavity and the degree of enantioselectivity of the interaction can both be altered by the nature of the surrounding solvent system. Cyclodextrins are often thought to act via the inclusion of a non polar portion of the guest into the hydrophobic cyclodextrin cavity. On the basis of this simple approximation we might expect any decrease in the hydrophobicity of the surrounding solvent to lead to a decrease in the degree of inclusion and vice versa. The solvent polarity is most obviously reduced by the addition of an organic solvent of a lower polarity than that of water. The enantioselectivity may also be altered by solvent polarity. Changes to the solvent polarity may alter the relative importance of the different interactions between the guest and the cyclodextrin (e.g. non polar, electrostatic and dipolar). If some of the interactions that are significantly altered by changing the solvent polarity are enantioselective ones then the ratio of non selective to selective interactions can also be altered.

Because changes to the solvent polarity can alter the affinity of the analyte for the cyclodextrin they can also change the optimum cyclodextrin concentration as was discussed in Chapter 3. As a result of it is not always easy to interpret the changes to enantioseparation with changes in solvent that are reported in the literature. Any change may be due to changes in enantioselectivity or to the chiral selector concentration being further from the optimum value.

It was shown that the resolution between the enantiomers of mandelic acid using γ-cyclodextrin in a phosphate buffer system at pH 7 could be either increased or decreased according to which organic solvent was added to the buffer. The addition of up to 20% acetonitrile gave a significant drop in resolution whereas the addition of up to 20% of either methanol or ethanol gave a large increase in resolution [17].

The enantioselectivity of β-cyclodextrin for the enantiomers of tioconazole was measured with different concentrations of organic solvents by Ferguson and others [18]. The enantioselectivity was measured in a triethanolamine-phosphate buffer system at pH 3.0 using acetonitrile concentrations of 0, 5, 10, and 15%, and with a citrate-phosphate buffer system at pH 4.3 using methanol concentrations of 0, 1, 4, 10, and 25%. Whilst the organic solvent decreased the binding constants by a factor of up to seven for acetonitrile and up to five for methanol, the enantioselectivity (as measured by the ratio of the equilibrium constants for the two enantiomers) was not altered. In this case the affinity of the analyte for the cyclodextrin was reduced without altering the relative importance of the enantioselective interactions.

In contrast to the result with tioconazole, other work showed that the enantioselectivity of a sulphated β-cyclodextrin for the enantiomers of terbutaline was strongly affected by the addition of organic solvents to the buffer system [19]. The enantioselectivity was measured in a phosphate buffer system in the absence of any organic solvent and with concentrations of methanol of 5 and 10%, and concentrations of acetonitrile of 5 and 10%. In both cases the addition of solvent lead to a decrease in the equilibrium constants but a marked increase in the enantioselectivity (ratio of the equilibrium constants). The increase in enantioselectivity is caused by the equilibrium constants for the more weakly binding enantiomer decreasing at a much faster rate than the more strongly binding enantiomer.

5.4.4 Thermodynamics of Inclusion

Temperature is an important experimental parameter in the analysis of achiral molecules by CE due to its influence for example on buffer viscosity and so migration times. With the separation of chiral molecules there is the additional interest in how the enantioselectivity varies as a function of temperature.

The mobile phase temperature is parameter which is often studied in HPLC be-

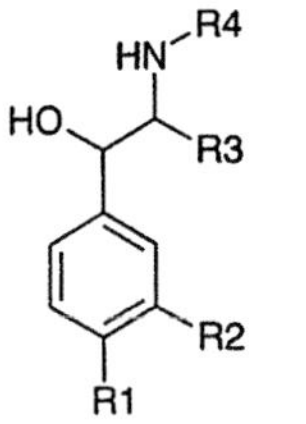

Figure 5.4. Biogenic amines used in thermodynamic studies.

Table 5.III.

Biogenic amine	R_1	R_2	R_3	R_4
β-Hydroxyphenethylamine	H	H	H	H
Norephedrine	H	H	CH_3	H
Ephedrine	H	H	CH_3	CH_3
Octopamine	OH	H	H	H
Synephrine	OH	H	H	CH_3
Norepinephrine	OH	OH	H	H
Epinephrine	OH	OH	H	CH_3
Isoproterenol	OH	OH	H	$CH(CH_3)_2$

cause of the resulting changes in enantioselectivity which can be observed. It is not surprising therefore to find that in CE the enantioselectivity may also depend upon the temperature of the buffer.

For achiral molecules the thermodynamics of inclusion have been studied by a number of methods such as microcalorimetry. The entropy and enthalpy changes are determined from the changes to the equilibrium constants with changing temperature using the van't Hoff isochore equation given in equation (5.1)

$$\ln K = \frac{-\Delta H}{RT} + \frac{\Delta S}{R} \qquad (5.1)$$

where K is the equilibrium constant, ΔH the enthalpy change, ΔS the entropy change, R the gas constant and T the absolute temperature.

The enthalpy and entropy changes upon the formation of complexes between α, β, and γ-cyclodextrins, and a large number of guest molecules have been reported [11]. In most cases the enthalpy change was negative and this was ascribed to the energy loss associated with the replacement of water from the hydrophobic cavity. The entropy changes were both positive and negative and varied with both the analyte and the cyclodextrin involved.

With enantiomeric molecules the interest is also in the differences in the thermodynamics of complexation between the two enantiomeric forms as described by equation (5.2).

$$\ln \left(\frac{K_2}{K_1} \right) = \frac{-\Delta \Delta H}{RT} + \frac{\Delta \Delta S}{R} \qquad (5.2)$$

where K_1 and K_2 are constants for the formation of the complexes between the cyclodextrin and the two enantiomers, $\Delta \Delta H$ is the enthalpy difference between the two complexes, $\Delta \Delta S$ is the entropy difference between the two complexes, R is the gas constant, and T is the absolute temperature.

There are several examples of literature CE studies showing the change in the separation of enantiomers with changing buffer temperature. It is not always apparent however whether the observed changes in separation are due to a change in the intrinsic enantioselectivity (e.g. the relative sizes of the complexation constants), or to a shift in the optimum chiral selector concentration. In Chapter 3 it was shown that the observation that enantioseparation could either increase or decrease upon the addition of an organic solvent could be explained by a simple equilibrium model. For some analytes there is an optimum concentration of chiral selector which is inversely related to the equilibrium constant for complexation as shown in equation (5.3).

$$[C]_{opt} = \frac{1}{\sqrt{K_1 K_2}} \qquad (5.3)$$

where $[C]_{opt}$ is the optimum chiral selector concentration and K_1 and K_2 are the formation constants for the two enantiomerchiral selector complexes.

The addition of an organic solvent might normally be expected to reduce K_1 and K_2 due to a decrease in the hydrophobicity of the cyclodextrin cavity relative to that of the surrounding buffer. Reduction in the equilibrium constants can result in either an increase or decrease in the effective mobility difference between the two enantiomers. The direction of change depends upon whether the new optimum chiral selector concentration (equation (5.3)) has moved closer to, or further away from the actual concentration.

As a change in temperature may also reduce the size of the equilibrium constants the same considerations should be made upon changing buffer temperature as are made upon changing buffer organic solvent content. A decrease in the observed enantioseparation with an increase in temperature may be due to a shift in the optimum chiral selector concentration away from the actual value, rather than a change in the intrinsic enantioselectivity.

The thermodynamics of both the inclusion and the enantioselectivity shown by β-cyclodextrin for ibuprofen enantiomers have been studied [20]. In a 10 mM sodium acetate system at pH 4.47 both the degree of inclusion (equilibrium constants) and the enantioselectivity (relative difference in the inclusion of the two enantiomers) appeared to decrease with increasing temperature. It was not possible to assign this decrease in enantioselectivity to differences between the enantiomers in either the entropy or enthalpy of inclusion ($\Delta \Delta S$ or $\Delta \Delta H$). The measured differences between the enantiomers are small relative to the uncertainties in measuring the overall enthalpy and entropy differences of inclusion (ΔH and ΔS).

The influence of temperature change on the degree of inclusion, and the enantioselectivity of inclusion, has been studied for the biogenic amines shown in Figure 5.4 and Table 5.III with dimethyl-β-cyclodextrin [21].

Both the degree of inclusion and the enantioselectivity of inclusion decreased with increasing temperature. The rate of change in the degree of inclusion with increasing temperature depended upon the structure of the analyte. The biggest changes were seen with the analytes having the most hydroxyl groups. The rate of change in enantioselectivity (as measured by the relative mobility difference between the two enantiomers) with temperature also depended upon the number of hydroxyl groups. The analytes with more hydroxyl groups were thought to form more hydrogen bonding interactions with the cyclodextrin and so are more strongly bound.

5.4.5 The Importance of pH

The pH is an important factor for consideration in the separation of enantiomers by CE as it determines the charges on both the analyte and the cyclodextrin. Altering the charge state of the analyte and chiral selector can affect both the degree of and the enantioselectivity of inclusion. The af-

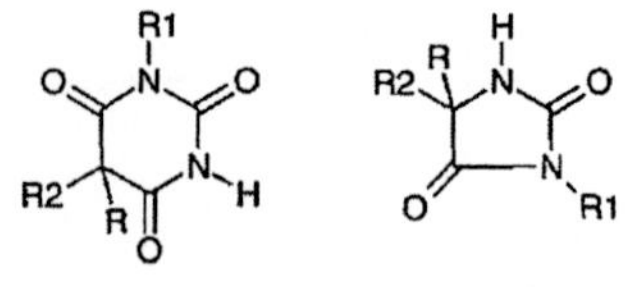

Figure 5.5. The structures of some barbituates and hydantoins.

Table 5.IV.

Compound	R_1	R_2	R	Name
4	CH_3	Phenyl	Ethyl	Methylphenobarbital
5	CH_3	Cyclohexen-1-yl	Methyl	Hexobarbital
6	CH_3	Cyclohexyl	Ethyl	
7	CH_3	Phenyl	Allyl	
8	H	Phenyl	Ethyl	
9	CH_3	Phenyl	Ethyl	Mephenytoin

finities of most analytes for the hydrophobic cavities of neutral cyclodextrins will decrease as the analytes become more highly charged and so less hydrophobic. Changes in pH which alter the charge on the analytes can therefore also alter the equilibrium constants for inclusion into a cyclodextrin cavity.

The enantioselectivity of a neutral cyclodextrin for enantiomers can also depend upon the charge on the enantiomers and three possible situations have been described by the theoretical models of Vigh et al. [22]. With *Type 1 enantiomers* the cyclodextrin only shows enantioselectivity for the neutral form of the enantiomers (*desionselective mechanism*), with *Type II enantiomers* the cyclodextrin only shows enantioselectivity for the charged forms of the enantiomers (*ionselective mechanism*), and with *Type III enantiomers* the cyclodextrin shows enantioselectivity for both the charged and neutral forms of the enantiomers (*duoselective mechanism*). The optimisation of pH is especially important when a neutral cyclodextrin only shows enantioselectivity for the neutral forms of the enantiomers. With a neutral cyclodextrin and a *desionselective mechanism* the pH chosen is a compromise between the need for neutral analytes to provide selectivity and charged analytes to give mobility difference.

Two example of a *desionselective mechanism* are those of the enantioselectivity of β-cyclodextrin for the enantiomers of both fenoprofen and ibuprofen [22, 20]. With fenoprofen the equilibrium constant for the both the charged forms is 325, and for the neutral *R* and *S* forms 608 and 636 respectively. With ibuprofen the equilibrium constant for the both the charged forms is 1280, and for the neutral *R* and *S* forms 1869 and 1954 respectively.

With achiral analytes it is known that the pK_a values of acids can be altered upon their complexation with cyclodextrins [3]. Complexation may make the acid stronger or weaker and the size of the shift in pK_a value depends upon the structure of the analyte. With 2-hydroxybenzoic acid for example the pK_a increased by 0.23

units whereas with 3-hydroxy benzoic acid the change was 0.85 units (p137 of [3]). It is possible therefore that with enantiomeric analytes different pK_a shifts will occur for the different enantiomers upon complexation. The possibility of different pK_a shifts for the two enantiomer-cyclodextrin complexes underlines the potential benefits of careful optimisation of both the chiral selector concentration and the pH.

5.4.6 Changes in Migration Order

Most major pharmaceutical companies have decided that new development compounds with a stereogenic centre should be produced as single enantiomers unless there are good reasons to do otherwise. Because of the decision to develop single enantiomers the newer pharmaceutical agents are usually very optically pure. The challenge for the analyst is therefore to measure the very low levels of the minor enantiomer (typically about 0.1% or less) in the presence of a large excess of the main enantiomer. Because of the poor concentration sensitivity of CE relative to HPLC it is usually necessary to overload the major enantiomer in order to see the minor one. The overloading leads to the peak for the main component either fronting or tailing as has been described in Chapter 4. Because of the tailing or fronting of the main peak there are potential benefits in resolution to be had from moving the minor enantiomer from one side of the major enantiomer to the other.

Changes in elution order are well known in the analysis of enantiomers by gas chromatography using different cyclodextrin bonded capillary columns. The elution orders of (−) and (+) isomenthol, and (+) and (−) α-benzylhexachloride change when the cyclodextrin stationary phase is changed from α-cyclodextrin to γ-cyclodextrin [23]. The nature of the cyclodextrin derivative in the stationary phase can also alter the enantioselectivity in gas chromatography. For example the

enantiomers of 2-butanol have different elution orders according to whether the stationary phase contains of the acetyl or the trichloroacetyl derivatives of γ-cyclodextrin [23].

Changes in elution order have also been observed in HPLC when cyclodextrins have been employed as mobile phase additives. The enantiomers of some of the barbituates and hydantoins shown in Figure 5.5 and Table 5.IV were separated by reversed phase HPLC using a C18 stationary phase and mobile phases containing either α, β, or Dimethyl-β-CD (DM-β-CD) [24].

With α, and β-cyclodextrin the retention of the analytes always decreased upon the addition of the cyclodextrin, and the decrease in retention was dependant upon the cyclodextrin concentration. The biggest decreases in retention were seen with β-cyclodextrin and in addition enantioselectivity was also observed for some of the analytes. This observation can be explained by competition between the hydrophobic sites on the stationary phase, and the hydrophobic cyclodextrin cavity for the enantiomers of the analytes. The reduction in retention was much higher for mobile phase containing β-cyclodextrin than α-cyclodextrin reflecting greater affinity. The enantiomer with the highest affinity for the cyclodextrin will spend the most time in the mobile phase and so will be eluted first.

With DM-β-CD as a mobile phase additive the situation was more complex. With analytes 4, 5, and 7 the retention *increased* at low DM-β-CD concentrations before decreasing at higher concentrations in the same way as for β-CD. In addition the elution order of the enantiomers of 4, 5, and 7, changed over with increasing concentration of DM-β-CD. The differences in the behaviour of DM-β-CD from that of β-CD were explained by the suggestion that DM-β-CD strongly adsorbed onto the stationary phase could also be playing a part in retention [25]. At low concentrations the cyclodextrin in the mobile phase would become adsorbed onto the stationary phase leading to an increase in retention. Once all of the adsorption sites were filled increasing the cyclodextrin concentration in the mobile phase would lead to an increasingly higher affinity for the mobile phase. The enantioselectivity at low concentration will be dominated by the stationary phase cyclodextrins, and at high concentration by the mobile phase cyclodextrins. Thus the enantiomer with the highest affinity for DM-β-CD would be the most retained at low concentration and the least at high concentration.

Changes in the migration order of the enantiomers with increasing cyclodextrin concentration have also been observed with electrically driven separations. For example the enantiomers of 1,1'-binaphthyl-2,2'-diylhydrogen phosphate were separated using a capillary coated with Chirasil-Dex with the (S)-enantiomer migrating the fastest [26]. The addition of increasing concentrations of sulphonated β-cyclodextrin to the buffer lead to a reduction in the observed enantioselectivity to zero. Further increases in the concentration of sulphonated β-cyclodextrin lead to a reversal in the migration order. When the Chirasil-Dex coated capillary was replaced with an uncoated capillary the (R)-enantiomer was the fastest migrating with sulphonated β-cyclodextrin in the buffer. The observed enantioselectivity for hexobarbital was also reduced when the same cyclodextrin was added to the buffer as was present in the capillary coating.

The reversal of migration order has been observed in a number of cases in CE and can be due to several different causes. The enantiomers of dansyl phenylalanine (Figure 5.6) were separated in a capillary coated with polyacrylamide using a pH 6 buffer containing hydroxypropyl-β-cyclodextrin (HP-β-CD) [27].

At a HP-β-CD concentration of 0.3% (w/v) the enantiomers were baseline resolved with the D enantiomer having the highest net mobility. At a concentration of 6% the enantiomers co-migrated, and at a HP-β-CD concentration of 15% the enantiomers were partly resolved with the L enantiomer having the highest net mobility.

Changes in migration order as a function of the chiral selector concentration can be explained by a theoretical model which is discussed more fully in Section 3.5.1. In essence enantioselectivity can arise both from the enantiomers having different affinities for the cyclodextrin, and/or the two enantiomer-cyclodextrin complexes having different limiting mobilities. If the charged enantiomer which binds most strongly to a neutral cyclodextrin also gives the complex with the higher limiting electrophoretic mobility then a change over in the migration orders is possible.

The CE literature contains many reports of different cyclodextrins giving different migration orders for a given enantiomeric pair. For example the migration order and degree of enantioselectivity for the enantiomers of some dansyl amino acids varied with both the position and degree of methylation on the β-cyclodextrin [28]. Methyl cyclodextrin methylated on the 3 position showed the opposite enantioselectivity for the dansylated derivatives of leucine, methionine, and norvaline to methyl cyclodextrin methylated on the 6 position. In another example the migration order of the enantiomers of naproxen (shown in Figure 5.7) was different with cyanoethylated β-cyclodextrin than with trimethyl-β-cyclodextrin [29].

Whilst differences in migration orders with different cyclodextrins have been reported on several occasions in the literature is not always clear whether these differences are due to differences in the enantioselectivity of binding, or due to differences in the limiting mobilities of the cyclodextrin-analyte complexes.

Changes to the migration order of the two enantiomers can also be achieved without changing the enantioselectivity of binding, by manipulation of the electroosmotic mobility and the electrophoretic mobilities of the analyte and chiral selector [30]. Changing from a neutral to a charged cyclodextrin means that, with charged analytes, the enantiomer which binds most strongly may have a higher rather than a lower effective mobility. Increasing the buffer pH can mean that the electroosmotic mobility is greater than the electrophoretic mobilities. So increasing the pH can mean that the charged enantiomer with the highest electrophoretic mobility becomes the enantiomer with the lowest apparent mobility. Changing from fused silica to coated capillaries can also change the size and direction of the electroosmotic mobility and hence the net mobility of a particular analyte.

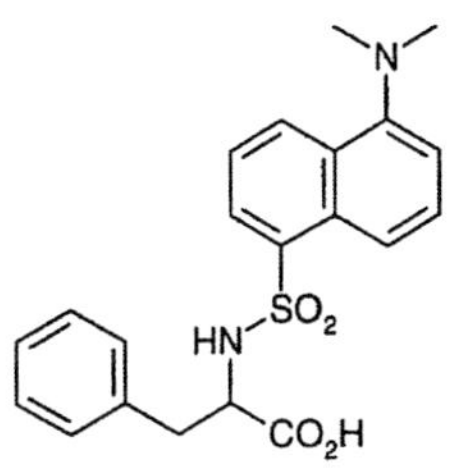

Figure 5.6. The structure of dansyl phenylalanine.

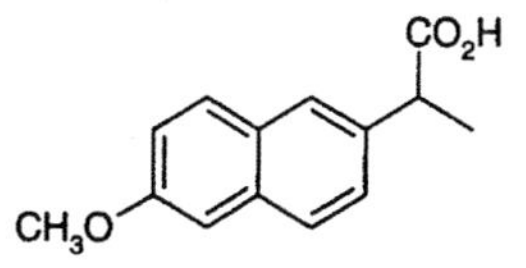

Figure 5.7. The structure of the anti inflammatory naproxen.

5.5 Cyclodextrin Classes

5.5.1 Neutral Cyclodextrins

Uncharged cyclodextrins were the first cyclodextrins to be used in the separation of enantiomers by CE and are probably still the most widely used. Most early work centred around the parent α, β, and γ-cyclodextrins but derivatised neutral cyclodextrins became popular very quickly. The attraction of many of the derivatised β-cyclodextrins over the parent is not only the differences in enantioselectivity, but also the much greater solubility in aqueous buffer systems.

The most commonly used neutral derivatives of β-cyclodextrin are the methyl, dimethyl, and trimethyl derivatives and the hydroxyethyl and hydroxypropyl derivatives. There are fewer articles citing the use of the neutral derivatives of α and γ-cyclodextrins but the hydroxypropyl derivatives have been used by several workers.

Predicting the best cyclodextrin is difficult with the present level of understanding of the factors affecting enantioselectivity and so the selection of the best cyclodextrin and the optimum cyclodextrin concentration is a largely matter of trial and error. There are a large number of examples in the literature of different enantiomers being resolved by the use of particular cyclodextrins, but it is not always clear whether the cyclodextrin used was the best of many examined or merely the first to give an adequate separation. The analyst can however draw upon some general observations on how the degree of inclusion can be related to the sizes of the analyte and cyclodextrin cavity. There are also some literature studies which have

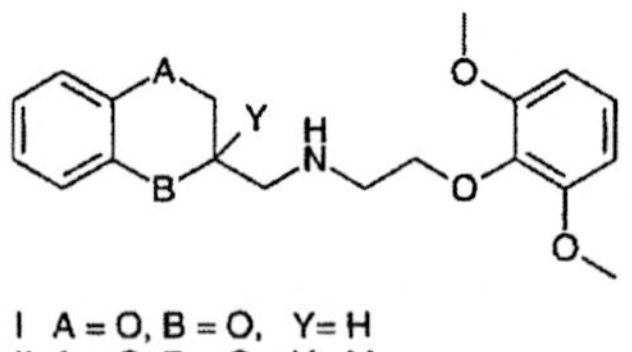

I   A = O, B = O,   Y = H
II A = O, B = O,   Y = Me
III A = O, B = CO, Y = H
IV A = S, B = O,   Y = H

Figure 5.8. The structures of some α_1 adrenoreceptor antagonists.

systematically evaluated different cyclodextrins.

Several groups have screened large numbers of analytes using a range cyclodextrins but often at only one or two concentrations. The use of a single or two similar concentrations can sometimes give misleading results as optimum cyclodextrin concentrations can vary widely according to the affinity of the analyte for the cyclodextrin. As some of the screening studies have, however, examined large numbers of analytes it is possible to draw some general conclusions about which cyclodextrins appear to give the greatest chance of success.

The resolution between the enantiomers of the α_1-adrenoreceptors shown in Figure 5.8 was measured using buffers containing either β-cyclodextrin or its dimethyl or hydroxypropyl derivatives at eleven different concentrations ranging from 1 mM to 60 mM [31].

The degree of resolution varied with both the cyclodextrin type and the cyclodextrin concentration. For analytes I, II, and IV the best resolution was obtained using hydroxypropyl-β-cyclodextrin and for analyte III the best resolution was obtained with β-cyclodextrin.

The resolution between the enantiomers of twenty basic pharmaceutical compounds including nine β-blockers was measured using buffers containing 15 mM of either β-cyclodextrin or its dimethyl, trimethyl, or hydroxypropyl derivatives [32]. The dimethyl and hydroxypropyl derivatives and the parent β-cyclodextrin showed enantioselectivity for a larger number of compounds than the trimethyl derivative. The trimethyl derivative gave some separation for only seven of the twenty compounds, whereas β-cyclodextrin and the dimethyl and hydroxypropyl derivatives gave partial or complete separation for thirteen and fourteen compounds respectively.

In another study the resolution between the enantiomers of twenty nine ba-sic and acidic pharmaceutical agents, and four dansylated amino acids was determined as part of a screening study employing γ-cyclodextrin, and β-cyclodextrins [33]. Each of the cyclodextrins was employed at two concentrations: β-cyclodextrin at 3 and 15 mM, γ-cyclodextrin at 10 and 50 mM, hydroxy-β-cyclodextrin at 10 and 100 mM, and dimethyl-β-cyclodextrin at 10 and 50 mM. Dimethyl-β-cyclodextrin gave the best resolution for seventeen of the analytes, hydroxypropyl-β-cyclodextrin gave the best resolution for ten of the analytes, β-cyclodextrin gave the best resolution for three of the analytes, and γ-cyclodextrin the best resolution for the remaining three analytes (all dansylated amino acids).

A similar study used the hydroxypropyl derivatives of α and γ-cyclodextrin, and the hydroxypropyl, dimethyl and sulphated derivatives of β-cyclodextrin to determine the resolution between the enantiomers of thirty eight commercially available basic drugs and proprietary pharmaceutical compounds [34]. The neutral cyclodextrins were used at two concentrations with both the hydroxypropyl, and dimethyl-β-cyclodextrins giving the best resolution on eleven occasions each, the hydroxypropyl-α-cyclodextrin on seven occasions, and the hydroxypropyl-γ-cyclodextrin on six occasions. Whilst the α and γ-cyclodextrins gave a lower level of success than the β-cyclodextrins they could be regarded as complementary and gave good resolution for some compounds which were only poorly separated by the β-cyclodextrins.

A further study used both parent α, β, and γ-cyclodextrins and the hydroxypropyl derivatives at a single fixed concentration (15 mM for β-cyclodextrin and 45 mM for the others) to measure the separation between the enantiomers of one hundred and twenty three pharmaceutical compounds [35]. Of the one hundred and twenty three compounds examined, eighty six were directly detectable by CE with the remainder being either insoluble or lacking a suitable UV chromophore. Sixty three of the eighty six compounds detected could be separated into their enantiomers including nine β-blockers and twenty three tricyclic compounds. All of the hydroxypropyl cyclodextrin derivatives gave a greater number of separations than their parents (approximately twice as many). The β-cyclodextrin produced the most separations amongst the native cyclodextrins, and hydroxypropyl-β-cyclo-dextrin performed better than the hydroxypropyl derivatives of α and γ-cyclodextrin.

5.5.2 Negatively Charged Cyclodextrins

In recent years charged cyclodextrins have become increasingly popular for the separation of enantiomers by CE. Charged cyclodextrins are required for the separation of the enantiomers of electrically neutral analytes, and are also beneficial for charged analytes – especially those carrying the opposite charge to that on the cyclodextrin. Cyclodextrins which can carry a negative charge due to the presence of an alkyl sulphonate derivative or other acidic group are especially important because of the large number of pharmaceuticals which contain a basic functional group.

Negatively charged cyclodextrins are useful for the separation of the enantiomers of positively charged drugs because they provide a means of maximising the electrophoretic mobility difference · between the free and complexed forms of the drug. In Chapter 3 a simple model was developed which described the apparent electrophoretic mobility difference between the two enantiomers and this is given below in equation (5.4)

$$\Delta\mu_{ep} = \frac{[C](\mu_0 - \mu_1)(K_2 - K_1)}{1 + [C](K_1 + K_2) + K_1 K_2 [C]^2}$$
(5.4)

where $\Delta\mu_{eph}$ is the apparent electrophoretic mobility difference between the two enantiomers; [C] is the chiral selector concentration; K_1 and K_2 the equilibrium constants for the formation of the two enantiomer-chiral selector complexes; μ_0 the electrophoretic mobility of the free enantiomers; and μ_1 the electrophoretic mobility of the complexed forms of the enantiomers.

From equation (5.4) it can be seen that for a given inherent enantioselectivity of binding (ratio of K_1 and K_2), the observed electrophoretic mobility difference will increase with the term $(\mu_0-\mu_1)$. This mobility difference can be maximised when the vector terms μ_0 and μ_1 have opposite signs, i.e. the bound forms of the enantiomers migrate in the opposite direction to that of the free forms. This mobility difference helps to explain the interest in cyclodextrins carrying several sulphonic acid or other acid side chains.

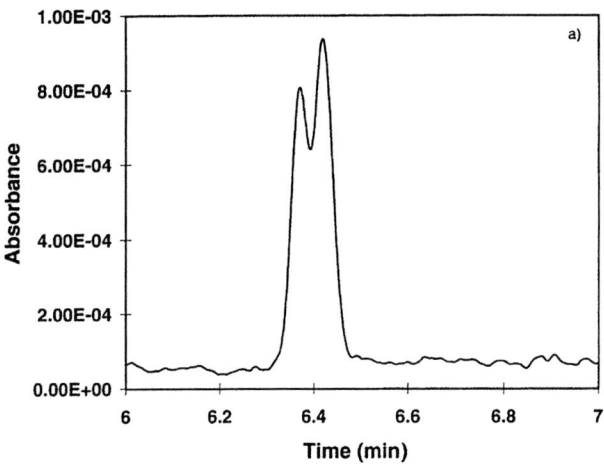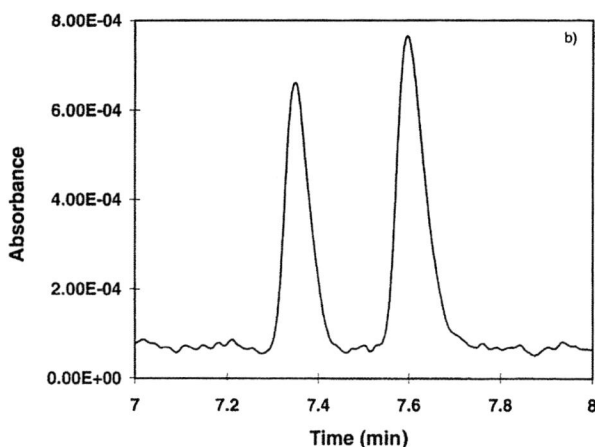

Figure 5.9. The resolution between the enantiomers of atenolol using **a)** dimethyl-β-cyclodextrin, and **b)** the sodium salt of sulphobutylether-β-cyclodextrin.

An example of the benefit of using a negatively charged cyclodextrin with a positively charged analyte is seen in Figure 5.9 with the superior resolution of the enantiomers of atenolol produced by using sulphobutylether-β-cyclodextrin rather than dimethyl-β-cyclodextrin [36]. The atenolol enantiomers were analysed in the same fused silica capillary using a lithium phosphate buffer at pH 2.2.

Figure 5.9a shows the resolution produced using 40 mM dimethyl-β-cyclodextrin and Figure 5.9b the resolution produced using 1.5 mM of the sodium salt of sulphobutylether-β-cyclodextrin. The optimum concentration of the charged cyclodextrin was much lower as the binding constants are much larger.

Cyclodextrins containing carboxylic acid groups were amongst the first charged cyclodextrins used in the separation of enantiomers. The carboxymethylated, carboxyethylated, and succinylated derivatives of β-cyclodextrin were used to resolve the enantiomers of doxylamine, ephedrine, dimetinden, and propranolol [37]. An interesting feature of carboxylic acid derivatives of cyclodextrins is that their charge can be modified according to the pH of the buffer system. The carboxymethyl derivative of β-cyclodextrin showed different selectivities at pH 2.5 (where it is electrically neutral) from those seen at pH 5.8 (where it is ionised). A comparison of the carboxymethyl and carboxyethyl-β-cyclodextrin derivatives indicated that the selectivities for different enantiomers depended upon the number of methylene groups in between the carboxylic acid group and the β-cyclodextrin.

Other workers have examined the use of cyclodextrins carrying a phosphate group [38, 39]. The phosphate derivatives of α, β, and γ-cyclodextrin were prepared by reacting the parent cyclodextrins with phosphorus oxychloride, and were used to separate the enantiomers of tocainide, metoprolol, and disopyramide [38]. The α and γ-cyclodextrin phosphates were found to give a higher selectivity for the enantiomers of tocainide than β-cyclodextrin phosphate.

Sulphonated cyclodextrin derivatives have been used by many workers for the separation of enantiomers, in particular those of bases. The use of sulphonic acid derivatives means that the derivatised cyclodextrins carry negative charge across a wide pH range. A sulphobutyl ether derivative of β-cyclodextrin was used to resolve the enantiomers of adrenaline, noradrenaline, ephedrine, pseudoephedrine, and tyrosine [40]. Beta-cyclodextrin was reacted with 1,4-butansultone to give a mixture of randomly substituted components with an average degree of substitution of four per cyclodextrin ring. The rationale for the alkyl chain in between the cyclodextrin cavity and the sulphonic acid group is to try and provide a spacing group which will allow the enantioselectivity of the cyclodextrin to remain unaltered. A comparison of the performance of the sulphobutylether-β-cyclodextrin with that of the parent and the dimethyl derivative showed the charged cyclodextrin to provide enantioselectivity at much lower concentrations. The sulphobutyl ether derivative of β-cyclodextrin was also used for the resolution of a range of basic and neutral pharmaceuticals, neutral phenyl alcohols, and dansylated amino acids [41].

The performance of charged and neutral β-cyclodextrins in the resolution of

Figure 5.10. The structure of (*S*)-duloxetine.

the enantiomers of eleven β-blockers was compared by using the carboxy methyl, sulphobutyl ether, dimethyl, and hydroxypropyl derivatives [42]. Two experimental designs were employed to screen conditions with three or four factors (cyclodextrin type, cyclodextrin concentration, buffer pH, and buffer methanol concentration) being examined at two or three levels respectively. The results showed that the negatively charged cyclodextrins were more effective chiral selectors for the β-blockers than the neutral cyclodextrins.

The degree of substitution of the sulphobutylether-β-cyclodextrin can also affect the observed enantioselectivity. A sample of mixed sulphobutylether-β-cyclodextrins was fractionated by ion exchange chromatography upon the basis of the number of charges, and the different fractions used to resolve the enantiomers of duloxetine (Figure 5.10) [43].

The fractions contained species with either 1, 2, 3, 4, 5, or 7 sulphobutylether units per β-cyclodextrin ring and the resolution of the duloxetine enantiomers produced by 0.1 mM solutions of the substituted cyclodextrins was measured. The observed resolution was seen to increase with the increasing degree of substitution although at a cost of increasing peak asymmetry.

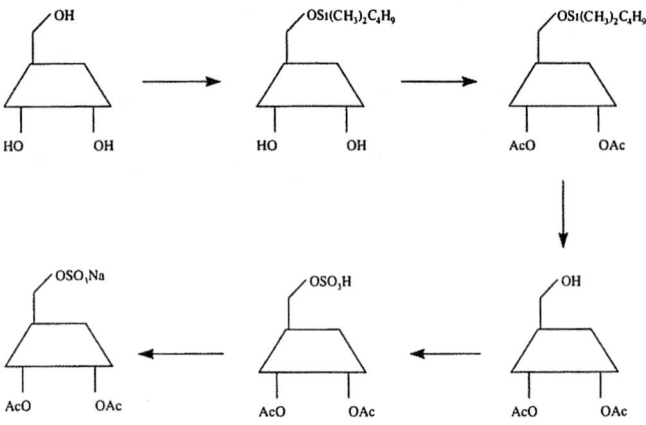

Figure 5.11. The synthesis of heptakis(2,3-diacetyl-6-sulphato)-β-cyclodextrin.

Fenoprofen

Indoprofen

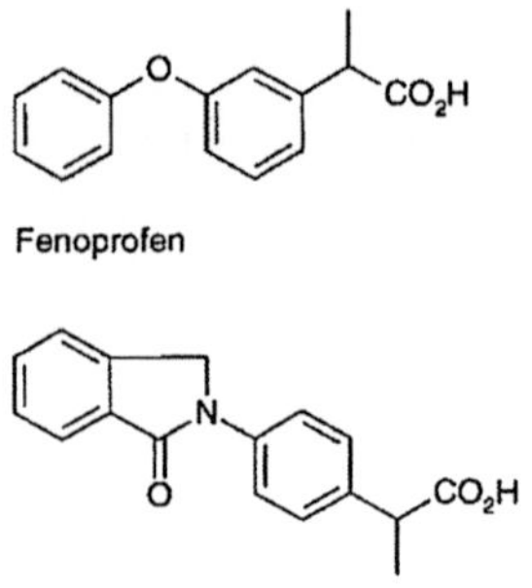

Figure 5.12. The non steroidal anti inflammatory drugs fenoprofen and indoprofen.

The influence of the degree of substitution on the observed enantioselectivity of different sulphobutyl ether derivatives of γ-cyclodextrin has also been investigated [44]. The different derivatives were produced by reacting γ-cyclodextrin with 1,4-butane sultone using different reaction times and temperatures and different ratios of the reactants. The average degree of substitution of the different mixtures was determined using MALDI-MS and elemental analysis. The different sulphobutylether-γ-cyclodextrin mixtures were used to resolve the enantiomers of binaphthol, secobarbital, 1-(9-anthryl)-2,2,2-trifluroethanol, and the dansylated derivatives of leucine, valine, and norvaline. The different derivatives were tested at the same molar concentrations and the resolution seen to vary according to the average degree of substitution.

The studies previously discussed show how the resolution varies according to the degree of substitution of the cyclodextrin. It is not certain however whether the differences in resolution arise from differences in the inherent enantioselectivity (K_1/K_2); to differences in the strength of binding; or to differences in electrophoretic mobility.

Sulphated cyclodextrins have also been successfully used for the resolution of the enantiomers of many pharmaceuticals. A sulphated β-cyclodextrin with a degree of substitution between seven and ten was used to resolve the enantiomers of 56 pharmaceuticals [45]. The pharmaceuticals analysed included anaesthetics, antihistamines, antiarrhymics, anticholinergics, antimalarials, antidepressants, β-blockers, and bronchodilators. The compounds were analysed in a 10 mM phosphate buffer at pH 3.8 using the sulphated β-cyclodextrin at a concentration of 2%.

Most of the resolved enantiomers were basic and many were very highly resolved with resolution values above three being common.

5.5.3 Single Isomer Negatively Charged Cyclodextrins

Most derivatisation procedures for cyclodextrins result in complex mixtures containing many species which vary both in the degree and position of substitution. This is problematic as the differently substituted species can have different enantioselectivities and different binding properties. These differences can lead to a number of concerns – e.g. that the future robustness of analytical control methodology may be compromised by changes in the quality of the chiral selector. In order to address this and other issues Vigh and co-workers developed a range of single isomer chiral selectors which are completely sulphonated at the 6 position and carry different functional groups at the 2 and 3 positions [46–49]. The key to the approach is the use of selective synthetic procedures which control the position and degree of substitution. Figure 5.11 for example shows the route used to produce heptakis (2,3-diacetyl-6-sulphato)-β-cyclodextrin.

Following the protection of the 6 position, different synthetic procedures were employed to produce heptakis (2,3-dimethyl-6-sulphato)-β-cyclodextrin, and hepta-6-sulphato-β-cyclodextrin. The different derivatives were synthesised to provide charged cyclodextrins with a range of different hydrophobicities. The single isomer sulphated cyclodextrins were successfully used to resolve the enantiomers of a wide range of compounds including neu-

tral, acidic, basic, and zwitterionic species. The hepta-6-sulphato-β-cyclodextrin was found to form the strongest complexes with the analytes and the dimethyl derivative the weakest.

5.5.4 Positively Charged Cyclodextrins

A wide range of positively charged cyclodextrins have also been used for the separation of enantiomers by capillary electrophoresis. Amino derivatives of β-cyclodextrin containing one or two aminomethyl groups have been used to separate the enantiomers of mandelic acid, some hydroxy derivatives of mandelic acid, and 2-, and 3-phenyllactic acid [50]. The enantiomers were separated using different β-cyclodextrins in phosphate buffers at pH 5 or 6, where both the analytes and the amino-β-cyclodextrins are charged. Both the degree of inclusion, and the degree of resolution was higher with the amino-β-cyclodextrins than with the parent β-cyclodextrin or 2-hydroxypropyl-β-cyclodextrin.

The enantiomers of seven Non Steroidal Anti Inflammatory Drugs (NSAIDs) and six phenoxypropionic acid herbicides were separated by heptakis (6-methoxyethylamine-6-deoxy)-β-cyclodextrin [51]. The separation between the enantiomers of the NSAIDs varied with the cyclodextrin concentration and pH. Resolution values of fourteen and eleven were obtained for the enantiomers of fenoprofen and indoprofen respectively (Figure 5.12). Changing the pH of the buffer altered the migration times of the analytes via the electroosmotic flow and the charges on the analytes and the cyclodextrin. The degree of resolution was strongly dependent upon the structure of the analyte with the maximum resolution of the enantiomers of ibuprofen being only one.

Quaternary ammonium derivatives of β-cyclodextrin, such as the 2-hydroxypropyltrimethylammonium salt have been used by several groups [52–54]. The 2-hydroxypropyltrimethylammonium derivative was prepared by reacting the parent β-cyclodextrin with 2,3-epoxypropyltrimethylammonium chloride as shown in Figure 5.13 [53].

This derivative was used to resolve enantiomers of some barbituates, warfarin, salbutamol, terbutaline, benzoin, thalidomide, and 5-methyl-5-phenylhydantoin [52, 53]. The use of polyacrylamide coated capillary was advantageous as with a fused silica capillary this quaternary salt reduced and then reversed the electroosmotic flow as the concentration was increased. With hexobarbital the peak symmetry was seen to improve with increasing concentrations of the quaternary chiral selector – presumably due to the increasing effective mobility reducing the effect of electrodispersion [52]. The resolution between the enantiomers of both basic and acidic analytes was shown to be a function of both the chiral selector concentration and the pH [54].

An interesting example of positively charged cyclodextrins are the two isomeric derivatives prepared from β-cyclodextrin and histamine [55]. The two possibilities involve the linkage to the cyclodextrin either through the amino group or through the imidazole nitrogen as seen in Figure 5.14.

These two histamine derivatives of β-cyclodextrin were used to resolve the enantiomers of ten dansylated amino acids, and found to give opposite enantioselectivities to each other. With the amino linked β-cyclodextrin derivative, the D-enantiomers of all of the amino acids except phenylalanine had the lowest migration times. With the imidazole linked β-cyclodextrin derivative, the L-enantiomers of all of the amino acids except phenylalanine had the shortest migration. The differences in enantioselectivities were ascribed to differences in the geometries of the histamine group, and the differences in the strength of the electrostatic interactions between the amino acid and the protonated amino and imidazole groups.

An interesting β-cyclodextrin derivative is that prepared with a glutamic acid side chain shown in Figure 5.15 [56].

Mono-(6-delta-glutamylamino-6-deoxy)-β-cyclodextrin (β-CD-Glu) is zwitterionic because of the presence of both amino and carboxylic acid groups and so may be either positively or negatively charged according to the pH. The derivative was used at both high and low pHs for the analysis chlorthalidone, benzoin and some of its derivatives, clenbuterol, carprofen, and flurbiprofen. The performance at pH 2.3 was also compared with that of amino-β-cyclodextrin. The data showed that whilst both cyclodextrins showed similar affinities for the neutral analytes (similar binding curves and optimum concentration), and similar selectivities, the resolution was about twice as high with the amino-β-cyclodextrin. The difference in resolution was ascribed to the greater mobility difference between the complexed and free forms of the analyte (equation (5.4)), caused by the greater charge on the amino-β-cyclodextrin. The amino-β-cyclodextrin is fully charged at pH 2.3 whereas β-CD-Glu has a net charge of about 0.5 because of the partial ionisation of the carboxylic acid group.

5.5.5 Mixed Selector Systems

There are many examples where the separation of enantiomers has been achieved or improved by the use of a cyclodextrin in conjunction with another selector species in the buffer. The additional selectors fall into two groups: achiral selectors such as the Sodium Dodecyl Sulphate (SDS); and a second chiral selector such as a chiral surfactant, or a second cyclodextrin. Mixed selector systems give greater flexibility although the greater complexity means that they are harder to model and may be harder to optimise. The achiral selectors may be used to generate electrophoretic mobility difference where neutral cyclodextrins have been used with neutral analytes. The use of a second chiral selector is more interesting and several groups have claimed that their mixed systems can give higher resolution than is possible than with either of the individual chiral selectors used in isolation.

Achiral selectors can be used with neutral cyclodextrins to either provide a means to generate an effective electrophoretic mobility difference between neutral enantiomers, or to improve enantioselectivity when either the analyte or cyclodextrin carries a charge. From equation (5.4) the observed mobility difference be-

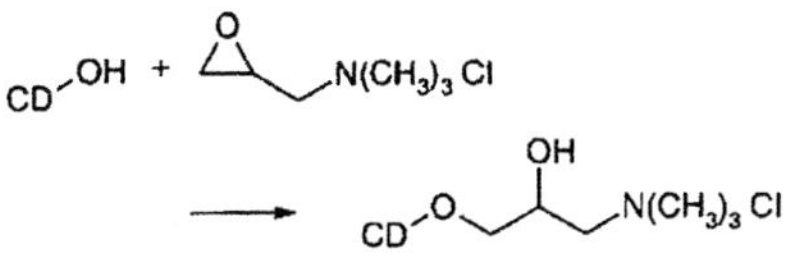

Figure 5.13. The synthesis of a quaternary ammonium cyclodextrin derivative.

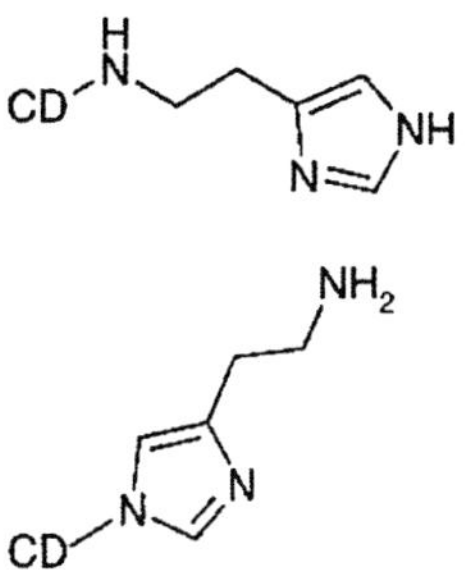

Figure 5.14. Isomeric derivatives formed by the reaction of β-cyclodextrin and histamine.

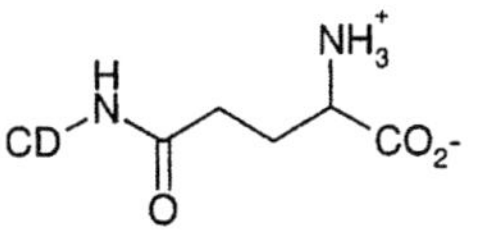

Figure 5.15. A glutamic acid derivative of β-cyclodextrin.

tween two enantiomers is seen to be proportional to the difference in electrophoretic mobility between the free and complexed form of the analyte. From equation (5.4) it follows that neutral enantiomers cannot be resolved by the use of a neutral chiral selector alone: regardless of the enantioselectivity of complexation, unless there are mobility differences between free and complexed forms of the analyte there will not be any separation. The use of a second selector carrying a charge means that the important competition process is not that between the free and complexed forms of the analyte, but between different complexed forms. Figure 5.16 shows the competition for the two enantiomers of a neutral analyte by a neutral cyclodextrin and a charged achiral surfactant such as SDS.

For example SDS was used in conjunction with either β or γ-cyclodextrin to separate the enantiomers of 2,2,2-trifluoro-1-(9-anthryl)ethanol [57].

Borate complexation has been employed to resolve the enantiomers of some neutral vicinal diols via the use of several β-cyclodextrin derivatives [58]. Borate ions can form complexes with diols having the correct geometry and the approach has also been used in the separation of sugars by capillary electrophoresis. Three of the enantiomeric analytes examined, phe-

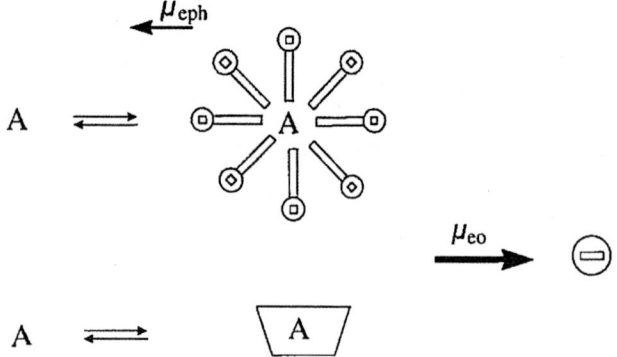

Figure 5.16. Enantioseparation of neutral analytes using a neutral cyclodextrin and a charged surfactant.

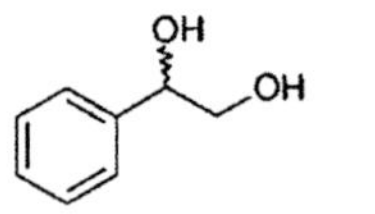

(R,S)phenyl-1,2-ethanediol

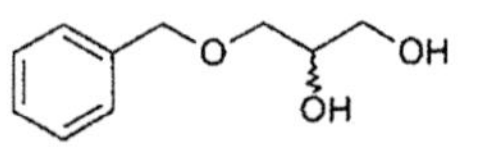

(R,S)3-benzyloxy-1,2-propanediol

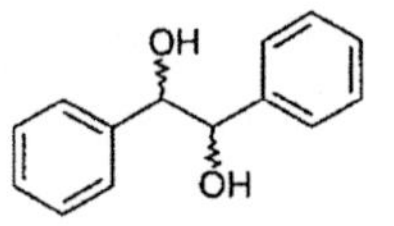

(RR)(SS)hydrobenzoin

Figure 5.17. The diols phenyl-1,2-ethandiol, benzyloxy-1,2-propanediol, and (RR), (SS) hydrobenzoin.

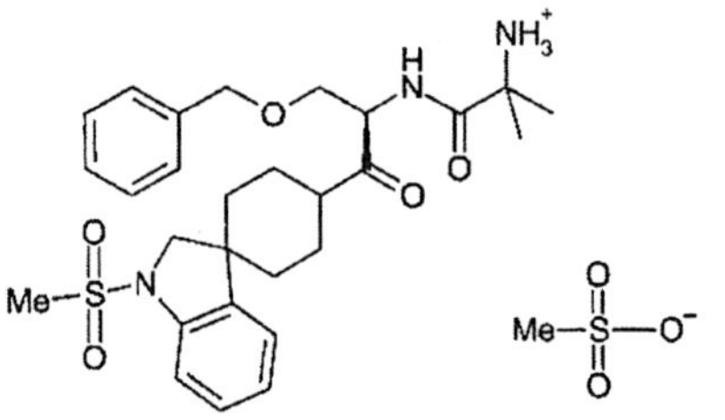

Figure 5.18. The structure of development compound MK-0677.

nyl-1,2-ethandiol and benzyloxy-1,2-propanediol, and (RR), (SS) hydrobenzoin are shown in Figure 5.17.

The use of 1.8% β-cyclodextrin and 50 mM borate at pH 9.3 gave resolution values ranging from 0.7 for the enantiomers of phenyl-1,2-ethandiol to 4.8 for the enantiomers of hydrobenzoin. The use of other buffer systems containing phosphate, TRIS, or CAPS did not lead to resolution of the enantiomers. The use of the di and trimethyl derivatives of β-cyclo-

dextrin did not give any resolution and this was thought to be due to the importance of borate complexation with the 2 and 3 hydroxyl groups on the cyclodextrin ring in addition to those on the analyte.

Small achiral additives in the buffer system have also been seen to improve enantioselectivity. The resolution between the enantiomers of propranolol using 10 mM β-cyclodextrin in a sodium phosphate buffer at pH 2.5 was improved by the addition of N(tert.-butoxycarbonyl)-glycine (TBCG) [59]. The resolution improvement was a function of the TBCG concentration with the optimum value being about 40 mM. Other additives such as tert.-butyl acetate, tert.-butyl carbamate, and tert.-butyl(N-hydroxyl)-carbamate at a concentration of 40 mM also lead to an improvement in resolution. Measurements indicated that these tert.-butyl additives reduced the degree of complexation of propranolol but made it more enantioselective. The addition of urea or ammonium carbamate to the buffer solutions containing β-cyclodextrin and TBCG lead to a complete loss of resolution. It is thought that the tert.-butyl additives may be incorporated into the β-cyclodextrin cavity and give a tighter and more stereoselective fit for the propranolol enantiomers.

The use of buffer systems containing a chiral selector in addition to the cyclodextrin have also been the subject of some interest. For example the resolution between the enantiomers of thiopental, pentobarbital, 2,2,2-trifluoro-1-(9-anthryl)ethanol, and 2,2'-dihydroxy-1,1'-dinaphthyl was improved by the use of d-camphor-10-sulphonate, and l-menthoxyacetic acid [57]. The two single enantiomer acids were added to a 20 mM phosphate/borate buffer at pH 9.0 containing 50 mM

SDS and 30 mM γ-cyclodextrin and shown to improve the resolution between the enantiomers of the analytes. With some analytes there seemed to be a continual change in resolution but in two cases with l-menthoxyacetic acid there seemed to be an optimum concentration.

Improved enantioselectivity was also observed by the combined use of β-cyclodextrin and the chiral surfactant taurodeoxycholic acid [60]. The resolution between the enantiomers of the dansylated amino acids threonine, valine, norvaline, tryptophan, glutamic acid, and aspartic acid were higher by the use of a combination of 20 mM β-cyclodextrin and 50 mM taurodeoxycholic acid, than by either selector used alone. The same dual system also gave superior resolution between the enantiomers of the antiepileptic drug mephenytoin, and those of its metabolite 4-hydroxymephenytoin. The interactions between the chiral surfactants sodium deoxycholate, and sodium cholate; and α, β, methyl β, hydroxypropyl, and γ-cyclodextrin have been studied by solution microcalorimetry [61]. The results enabled the estimation of thermodynamic parameters and showed the formation of 1:1 complexes between the chiral surfactants and the cyclodextrins.

Other groups have studied the enhancements to the enantioselectivity of cyclodextrin that can be obtained by the use of chiral ion pair reagents. Some of the data indicate that the chiral ion pair reagents are not necessary and that equivalent results can be achieved from the use of achiral ion pair reagents. For example the enantiomers of the pharmaceutical development compound MK-0677, Figure 5.18, were resolved using a buffer system containing 30 mM β-cyclodextrin and 40 mM L-tartaric acid in a phosphate buffer at pH 4.2 [62].

The resolution was a function of the concentration of L-tartaric acid with the data indicating an optimum concentration. The resolution seems to partly depend upon the competition between the concentrations of the β-cyclodextrin and L-tartaric acid. The L-tartaric acid appears to reduce the affinity of the analyte for the cyclodextrin, and so can make the cyclodextrin concentration either closer to or further away from the optimum value. Similar changes to the resolution could be obtained from the use of D-tartaric acid, camphorsulphonic acid, and achiral acids such as citric acid and 3-(N-morpholino)-propansulphonic acid. Measurement of the equilibrium constants for complexation indicated that the acids reduced the degree of binding but also increased the enantioselectivity (ratio of the equilibrium constants).

Another larger study examined the affect of acidic and basic ion pair reagents on the resolution between the enantiomers of some basic and acidic analytes produced using β-, methyl β-, hydroxypropyl β-, and γ-cyclodextrin [63]. The analytes included the bases cyclodrine, prilocaine, bupivacaine, disopyramide, homatropine, and terbutaline; and the acids included phenylacetic acid and homologs, and mandelic acid and some related species. The basic analytes were analysed in phosphate buffers at pH 2 or 7 using 1.8% of the cyclodextrin, in both the presence and absence of 40 mM of the ion pair reagents including camphorsulphonic acid, various alkane sulphonic acids, and various alkanoic acids. The acidic ion pair reagents could either increase or decrease the separation of the basic analytes and it was difficult to see any clear patterns. The acidic analytes were analysed in an acetate buffer at pH 5 using 15 mM of the cyclodextrin in both the presence and absence of 40 mM quinine. The addition of quinine gave an increase in the separation factor with some analytes and using some cyclodextrins and a decrease with others. With methyl and hydroxypropyl-β-cyclodextrin the quinine almost always lead to a decrease or no change in separation whereas with γ-cyclodextrin the separation was always as good or better. It is possible that the ion pair agents are acting in two ways: 1) by reducing the relative affinity of the analyte for the cyclodextrin (and so making the cyclodextrin concentration closer to or further away from the optimum value); and 2) by lining the cyclodextrin cavity and so changing the degree of fit of the analyte.

5.5.6 Dual Cyclodextrin Selector Systems

The use of buffer systems containing more than one cyclodextrin have become popular in recent years for the separation of enantiomers. Several groups have shown better enantioselectivity with dual cyclodextrin systems than that obtained from the use of either of the cyclodextrins alone. In most cases a charged cyclodextrin is used in conjunction with a neutral cyclodextrin. With some of the neutral

analytes examined the charged cyclodextrin is thought to act merely as a carrier for the neutral analyte. In these cases the charged cyclodextrin shows no enantioselectivity itself, but provides a mechanism whereby the enantioselectivity of the neutral cyclodextrin can be expressed. In other cases with charged analytes both cyclodextrins can contribute to enantioselectivity. Alternatively one cyclodextrin may be responsible for another form of discrimination such as that between diastereoisomers or related substances.

The resolution between the enantiomers of several non-steroidal anti inflammatory drugs (NSAIDs) such as ibuprofen and fenoprofen (Figure 5.19) was superior using a combination of sulphobutylether-β-cyclodextrin and trimethyl-β-cyclodextrin than either of the cyclodextrins used in isolation [64].

The analysis of the NSAIDs was performed in a triethanolamine-phosphate buffer at pH 3, meaning that the analytes were almost fully protonated. At pH 3 the use of neutral cyclodextrins alone could not be expected to lead to any resolution because of the lack of any meaningful mobility difference between the free analytes and those complexed to the cyclodextrin. With the addition of 5 mM sulphobutylether-β-cyclodextrin (with an average degree of substitution of four per cyclodextrin ring) some resolution was observed between the enantiomers of some of the analytes. The use of the charged cyclodextrin gave a resolution of < 0.7 for the enantiomers of fenoprofen, and no resolution between the enantiomers of ibuprofen (although the migration time change indicated a degree of complexation). The combination of 10 mM of trimethyl-β-cyclodextrin and 5 mM sulphobutylether-β-cyclodextrin gave a resolution of 2.8 between the enantiomers of fenoprofen, and 1.5 between the enantiomers of ibuprofen.

The use of a combination of a neutral and a charged cyclodextrin was also beneficial in the resolution of the enantiomers of a range of barbituate drugs including pentobarbital and secobarbital (Figure 5.20) [65].

The resolution between the enantiomers of pentobarbital and secobarbital was measured in a 100 mM triethanolamine-phosphate buffer at pH 3 using a range of concentrations of carboxymethyl-β-cyclodextrin (CMBCD) between 1 and 15 mM. For each of the analytes the maximum resolution value obtained was about 1.1 at an optimum CMBCD con-

Figure 5.19. The NSAIDs fenoprofen and ibuprofen.

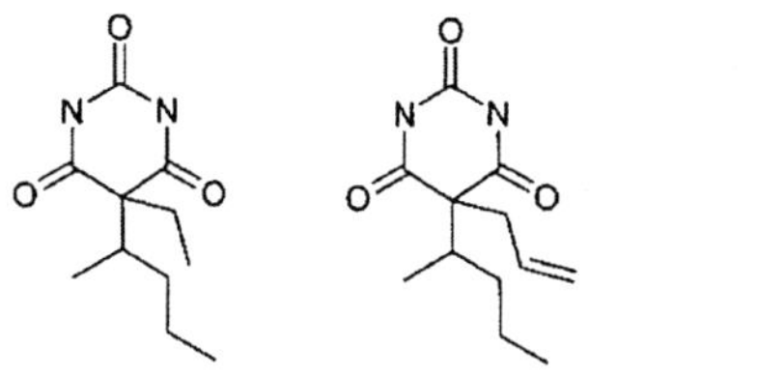

Figure 5.20. Pentobarbital and secobarbital.

Figure 5.21. Venlafaxine and its metaboloite O-desmethylvenlafaxine.

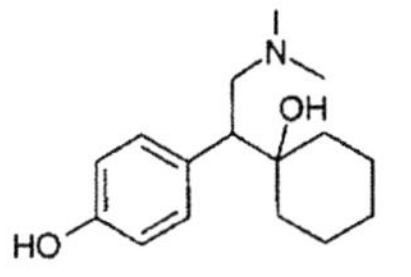

centration of 10 mM. The addition of 10 mM of trimethyl-β-cyclodextrin lead to an increase in the resolution values to 2.4 for pentobarbital and 1.6 for secobarbital respectively. The resolution between the enantiomers of a range of analytes was measured using charged β-cyclodextrins alone or in combination with neutral β-cyclodextrins. The resolution was usually superior using a combination of carboxymethyl-β-cyclodextrin or sulphobutylether-β-cyclodextrin, and di- or trimethyl-β-cyclodextrin. The trimethyl-β-cyclodextrin was particularly effective with the combination approach leading to an improvement in resolution in eleven out of twelve cases (in the twelfth case there was no change to the resolution).

The second cyclodextrin can also be useful in the separation of related substances. For example the enantiomers of venlafaxine (Figure 5.21) were resolved from each other, and from the enantiomers of its main metabolite O-desmethyl-

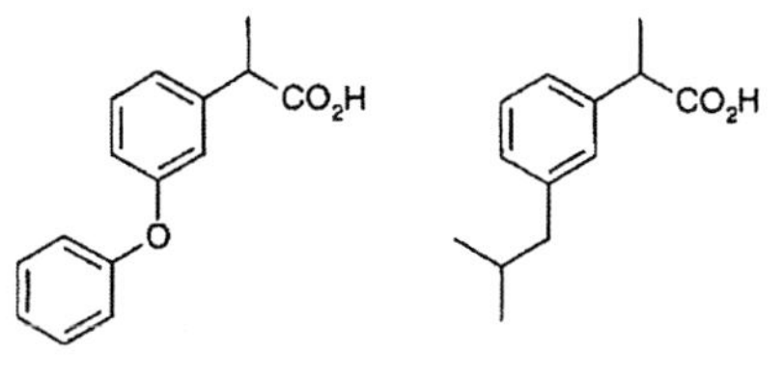

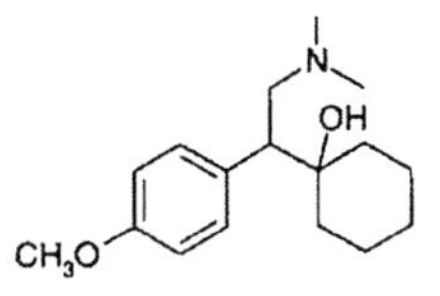

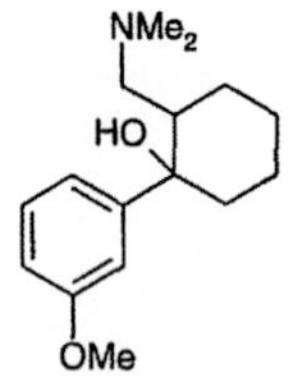

Figure 5.22. The structure of tramadol.

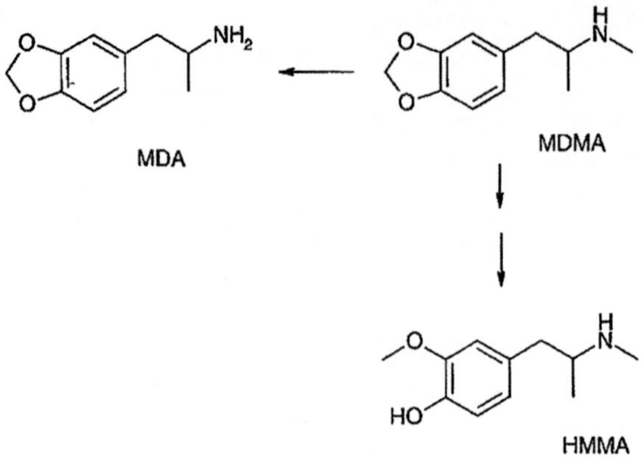

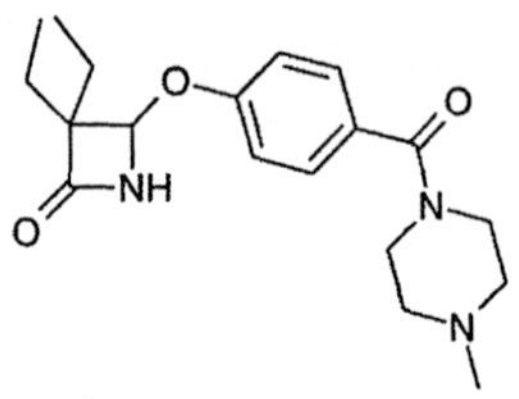

Figure 5.24. The β-lactam piperazinyl amide intermediate.

Figure 5.23. 3,4-methylenedioxymethamphetamine and some of its metabolites.

velafaxine by the use of a buffer system containing both carboxymethyl-β-cyclodextrin and α-cyclodextrin [66].

The use of carboxymethyl-β-cyclodextrin by itself gave good resolution between the enantiomers of the two chemical components but their was partial overlap between one of the enantiomers of venlafaxine and one of the enantiomers of O-desmethylvenlafaxine. The addition of α-cyclodextrin ensured that four peaks due to the two chemical components and their enantiomers were fully resolved.

5.6 Analyte Classes

The majority of the literature work on the separation of enantiomers by capillary electrophoresis using cyclodextrins has been directed towards the analysis of the active agent of ethical pharmaceuticals. Enantiomer separations by CE are also important in a number of other areas and cyclodextrins have also been used for example in metabolism studies, process analysis, forensic analysis of drugs of abuse, natural products, and agrochemical agents such as herbicides and pesticides. The following applications are examples only and are by no means exhaustive.

5.6.1 Metabolism Studies

Sulphobutylether-β-cyclodextrin was used to separate the enantiomers of the analgesic tramadol (Figure 5.22) and those of its mono, di, and tri desmethyl metabolites [67]. Electrospray mass spectrometry was used as the detection technique and the use of single ion monitoring helped to distin-

guish between the different possible metabolites. The analysis of the plasma sample from a volunteer showed the presence of the two enantiomers of tramadol in addition to those of the O-desmethyl metabolite.

The 2-hydroxypropyl derivative of β-cyclodextrin was used to determine the metabolic fate of the enantiomers of 3,4-methylenedioxymethamphetamine (MDMA), a stimulant and euphoric commonly known as ecstasy [68]. The levels of the enantiomers of MDMA and those of two of its metabolites, 3,4-methylenedioxyamphetamine (MDA) and 4-hydroxy-3-methoxymethamphetamine (HMMA) (Figure 5.23), were measured in the urine of two human patients during the three days following oral administration of racemic MDMA.

The metabolism of MDMA was found to be stereoselective with significantly higher amounts of (R)-(−)-MDMA than (S)-(+)-MDMA being excreted in the three day period. Whilst the majority of the MDMA was excreted unchanged the metabolites MDA and HMMA were also detected with different levels of the enantiomers being measured. The data also showed differences in the metabolism of the two patients.

5.6.2 Process Analysis

Cyclodextrins have also been used to monitor the enantiomeric purity of intermediates used in the synthesis of single enantiomer pharmaceutical agents. An example is the use of a buffer containing 5 mM β-cyclodextrin and 100 mM sodium phosphate at pH 2.5 to separate the enan-

tiomers of the β-lactam piperazinyl amide shown in Figure 5.24 [69]. The desired enantiomer is (S) and this was produced by a classical resolution of a racemic starting material using L-gulonic acid. The (R) enantiomer precipitates from solution leaving the required (S) enantiomer enriched in the mother liquors. The unwanted (R) enantiomer is recycled in the process by conversion to the racemate. CE was used to measure the levels of the two enantiomers in the precipitate and the mother liquors and was used to optimise both the enantiomeric and chemical purity.

5.6.3 Forensic Analysis

Cyclodextrins have also been used in capillary electrophoresis for the analysis of enantiomers of forensic interest. Dimethyl-β-cyclodextrin and sulphobutylether-β-cyclodextrin have both been used for the analysis of the enantiomers of the components contained within illicit drugs such as cocaine, khat leaves, and amphetamines [70]. In several cases only one enantiomer is illegal and so enforcement agencies must demonstrate that the correct enantiomer is involved. The determination of the enantiomeric purity also helps to indicate the source of the illicit drugs and also the age of the sample. For example the presence of racemic amphetamine suggests synthesis from an achiral starting material such as phenylacetone, whereas a single enantiomer suggests stereoselective reduction of norephedrine or norpseudoephedrine (Figure 5.25).

A triethylamine phosphate buffer was used in combination with either (or both) 5 mM dimethyl-β-cyclodextrin or sulphobutyl-ether-β-cyclodextrin to separate the enantiomers of a range of components of interest such as cathinone, amphetamine, methamphetamine, ephedrine, and cocaine.

5.6.4 Carbohydrates and Peptides

The enantiomers of some derivatised monosaccharides were separated using a borate buffer system containing 15 mM β-cyclodextrin and 80 mM taurodeoxycholate [71]. The derivatives were formed by reacting the D- and L-enantiomers of galactose, fucose, and ribose with 2-aminoacridone followed by the reduction of the Schiff base intermediates. The derivatives of the monosaccharide enantiomers were detected by laser induced fluorescence using an argon laser to excite at 488 nm with measurement at 520 nm.

The enantiomers and diastereoisomers of some di and tri amino acid peptides have been separated using some negatively charged cyclodextrins. Sulphobutyl-ether-β-cyclodextrin was used in conjunction with some neutral surfactants to separate an acylated Asp-Phe dipeptide drug candidate from its enantiomer and diastereoisomers [72]. The diastereoisomers included those formed from the isomerisation reaction involving the aspartic acid residue.

Sulphonated derivatives of α, β, and γ-cyclodextrin were used to separate the enantiomers and diastereoisomers of some aspartyl di- and tripeptides [73]. The enantiomers and diastereoisomers of Asp-Phe-OMe, Asp-Phe-NH$_2$, and Gly-Asp-Phe-NH$_2$ included those arising from isomerisation of the aspartic acid group (e.g. Figure 5.26).

All of the eight species arising from Asp-Phe-NH$_2$ were baseline separated using 5% w/v of highly sulphonated α-cyclodextrin in a pH 3 buffer. For all of the peptides the resolution was found to vary according to the size of the cyclodextrin, the degree of substitution and the buffer pH.

5.6.5 Alkaloids

Cyclodextrins have been used to separate the enantiomers of some important alkaloids. For example the enantiomers of the ergot alkaloids isolysergic acid, terguride, meluol, nicergoline, and lisuride were se-

parated using γ-cyclodextrin in a 100 mM phosphate buffer at pH 2.5 [74].

Cyclodextrins have also been used to analyse the enantiomeric purity of atropine in pharmaceutical preparations and vegetable extracts. Atropine, or dl hyoscyamine (Figure 5.27), is one of the principle alkaloids found in the solanaceae family of plants which includes deadly nightshade. The Latin name for deadly nightshade, Atropa bella-donna, is thought to derive from the former usage of the plant by women to dilate their pupils and so appear more attractive. Pharmaceutical grade atropine is an equal mixture of d and l hyoscyamine due to the racemisation which occurs during extraction and purification.

Trimethyl-β-cyclodextrin was used in a 100 mM phosphate buffer at pH 2.5 to separate the hyoscyamine enantiomers and measure their levels in some pharmaceutical preparations and some crude drugs prepared from extracts of the root Scopolia japonica (a member of the solanaceae family) [75]. The two pharmaceutical preparations (eye drops and an injection) were shown to contain equal levels of the two enantiomers. The different Scopolia extracts showed a predominance of the l hyoscyamine enantiomer although the levels varied between 74% and 90% in the different samples. These differences in enantiomeric purity are probably a reflection of differences in the extraction procedures and may be significant as l hyoscyamine is more active than d hyoscyamine.

The hyoscyamine enantiomers were also resolved using a phosphate buffer system at pH 7 containing sulphated β-cyclodextrin [76]. These analytical conditions also allowed the resolution of littorine, a closely related positional isomer of hyoscyamine. The optimised separation conditions were obtained by the use of a central composite experimental design to vary the buffer pH and concentration, and the concentration of sulphated β-cyclodextrin. The analytical method was used to measure the levels of the hyoscyamine enantiomers in an ophthalmic solution and some extracts prepared from the roots of the plant Hyoscyamus albus. The ophthalmic solution was shown to contain equal amounts of the two enantiomers but the Hyoscyamus albus extracts contained much higher levels of l hyoscyamine (92–98%). The extract prepared by supercritical fluid extraction showed less racemisation than the extracts prepared by liquid-solid extraction.

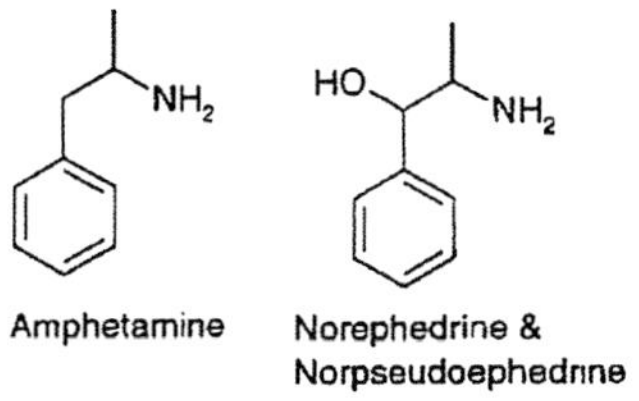

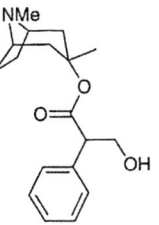

Amphetamine   Norephedrine & Norpseudoephedrine

Figure 5.25. Amphetamine, Norephedrine, and Norpseudoephedrine.

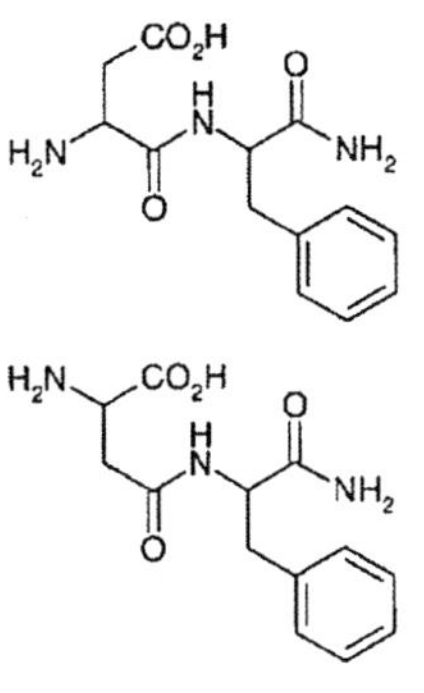

Figure 5.26. α- and β-Asp-Phe-NH$_2$.

Figure 5.27. Hyoscyamine (atropine).

5.6.6 Herbicides

Cyclodextrins have been used for the separation of the enantiomers of the phenoxy acid herbicides dichloroprop, mecoprop, and fenoprop shown in Figure 5.28 [77]. Enantiomeric separation is important as the herbicidal activity of dichloroprop and mecoprop resides in the d enantiomer.

The enantiomers of the individual herbicides were baseline resolved from each other, and those of the other herbicides, by the use of a 25 mM solution of trimethyl β-cyclodextrin in a 50 mM sodium acetate buffer at pH 4.45. A range of buffer systems of different types and pHs and concentration were examined but the 50 mM sodium acetate buffer at pH 4.45 gave the best separation between the different herbicides. Several other cyclodextrins were examined but only the trimethyl-β-cyclodextrin gave satisfactory resolution between all pairs of enantiomers.

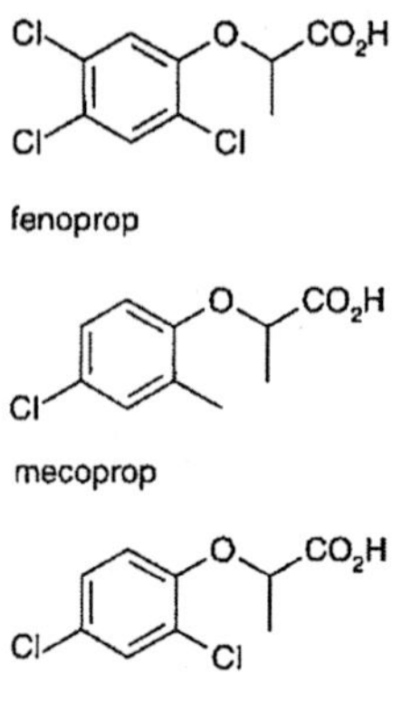

fenoprop

mecoprop

dichlorprop

Figure 5.28. The phenoxy acid herbicides fenoprop, mecoprop, and dichloroprop.

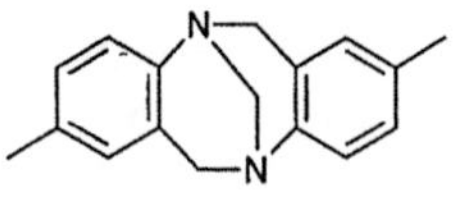

Figure 5.29. Tröger's base.

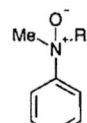

R = Ethyl, *n*-Propyl, *i*-Propyl, *n*-Butyl, *i*-Butyl
CH₂CCH

Figure 5.30. Enantiomeric *N*-oxides.

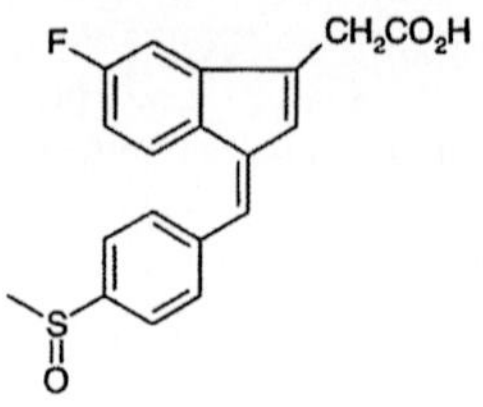

Figure 5.31. Sulindac.

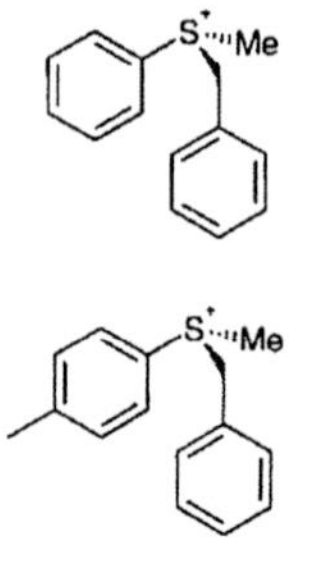

Figure 5.32. Benzylmethylphenyl-, and benzyl-methy-*p*-tolylsulphonium ions.

5.7 Analyte Structure

Most of the CE literature on the separation of enantiomers using cyclodextrins relates to cases where the enantiomers arise from a tetrahedral carbon atom with four different substituent groups. The following cases are examples where carbon is not the atom involved at the stereogenic centre or where the enantiomers arise from another property such as hindered rotation.

5.7.1 Stereogenic Nitrogen Atoms

Pyramidal nitrogen compounds containing three different substituents do in principle have a stereogenic centre with the nitrogen lone pair acting as the fourth ligand. In most cases involving pyramidal nitrogen compounds the energy barrier to inversion is, however, low and so the enantiomeric forms cannot be separated. There are some exceptions to rapid inversion at nitrogen centres in cases where the three ligands are joined together in a cage structure. A notable example of a rigid structure preventing inversion at nitrogen is that of Tröger's base shown in Figure 5.29. The enantiomers of Tröger's base were easily resolved by the use of a sulphated cyclodextrin in a sodium phosphate buffer at pH 3.8 [45].

The oxidation of tertiary amines having three different substituents results in the formation of enantiomeric amine *N*-oxides. Some of the enantiomeric *N*-oxi-

des have been separated by CE using cyclodextrins. For example β-cyclodextrin and its methyl, hydroxyethyl and hydroxypropyl derivatives were used to separate the enantiomers of some *N*-alkyl-*N*-methylbenzylamine *N*-oxides, and those of pargyline *N*-oxide (Figure 5.30) [78].

The enantiomeric *N*-oxides were separated in 150 mM or 300 mM phosphate buffer at pH 2.5. Whilst none of the cyclodextrins proved suitable in all cases, all of the enantiomers could be satisfactorily resolved by optimisation of the cyclodextrin type and concentration and the concentration of the buffer. A buffer containing 50 mM of hydroxyethyl-β-cyclodextrin for example was used to separate the enantiomers of pargyline *N*-oxide. The migration order of the enantiomers of pargyline *N*-oxide was established by spiking experiments and the degree of resolution meant that the levels of minor enantiomer could be detected below the 1% level.

The CE method developed was used to determine the stereoselectivity of the oxidation of pargyline by different forms of the enzyme flavin-containing monooxygenase (FMO). The stereoselectivity and activity of two different isoforms of recombinant human FMO, FMO1 and FMO3, was determined. FMO1 was found to be nearly stereospecific for *d* pargyline *N*-oxide and the less active FMO2 to be stereoselective for *l* pargyline *N*-oxide.

5.7.2 Stereogenic Sulphur Atoms

There are several examples of the use of cyclodextrins in CE to separate enantiomers of compounds with sulphur at the stereogenic centre. The enantiomers of the non steroidal anti inflammatory drug sulindac (Figure 5.31) were separated by the use of sulphobutylether-β-cyclodextrin [64]. The sulphobutylether-β-cyclodextrin was used either alone or in combination with β-cyclodextrin or neutral derivatives such as dimethyl, trimethyl, and hydroxypropyl-β-cyclodextrin.

There are also several examples of the use of cyclodextrins to separate enantiomers of enantiomeric sulphonium ions. Trivalent pyramidal derivatives of sulphur often have the potential to be resolved into their enantiomers as the energy barrier to inversion is sufficiently large to limit the rate of interconversion under normal conditions. The enantiomers of a range of sulphonium ions such as the benzylmethylphenylsulphonium, and benzylmethyl-p-tolylsulphonium ions (Figure 5.32) were resolved by the use of β-cyclodextrin in a sodium phosphate buffer at pH 2.5 [79]. The resolution depended on the cyclodextrin concentration and was found to improve by the addition of tetrabutylammonium bromide to the buffer. The related sulphonium ions formed by replacement of the methyl group with another alkyl group could also be separated into their enantiomers.

The sulphonium ions produced by the replacement of the benzylmethyl group with a second alkyl group could not however be separated into their enantiomers by the use of β-cyclodextrin. These methyl alkyl sulphonium ions could only be resolved by the replacement of β-cyclodextrin with its sulfated derivative. The resolution between the enantiomers of the methyl alkyl sulphonium ions was found to depend upon the length of the alkyl chain with the propyl species being the most poorly resolved.

5.7.3 Hindered Rotation

Enantiomers (atropisomers) can also be generated in cases where hindered rotation leads to molecules which are normally planar adopting a twisted conformation. An example of a molecule with hindered rotation is that of 1,1'-binaphthyl-2,2'-diol, which has been separated into its enantiomers by several workers using cyclodextrins and other selector systems. Another important example of molecules with hindered rotation being separated into enantiomers by the use of cyclodextrins is that seen with some of the polychlorinated biphenyls. Chlorination on the 6 (or 6 and 6') and 2 and 2' positions on the biphenyl ring leads to sufficiently hindered rotation and so the possibility of separation of the enantiomers. For example the enantiomers of 2,2',3,4,6-pentachlorobiphenyl were baseline separated by the use of a dual buffer system containing both γ-cyclodextrin and bile salts [80].

5.8 Batch and Source Variation

Most commercially available substituted cyclodextrins contain a mixture of components which differ in both the degree and position of substitution. For example whilst most dimethyl-β-cyclodextrins may be described as "heptakis(2,6-di-O-methyl)-β-cyclodextrin" the 3 position may also be partially methylated and the 2 and 6 positions may be incompletely methylated. The importance of major differences in the methylation pattern was investigated by selectively synthesising the different mono and dimethyl substituted β-cyclodextrins and using them to resolve the enantiomers of some dansyl derivatised amino acids [28]. The different methyl and dimethyl derivatives gave differences in resolution and in some cases migration order.

As the enantioselectivity of the components in a cyclodextrin mixture can differ, the batch to batch and source to source reproducibility of the cyclodextrin can be an important issue. Batch to batch and source to source variation may be an especially important issue where the CE enantioseparation method is intended to be used over a long period of time and should be considered as part of method validation.

The performance of seven different methylated β-cyclodextrins was determined by comparing the degree of resolution that they produced between the enantiomers of N-methylephedrine, terbutaline and hexobarbital [81]. The cyclodextrin samples included those which were produced as dimethyl derivatives and those which contained a more random distribution. The degree of resolution varied between the different cyclodextrins with the degree of variability was related to the analyte. The resolution varied between 1.43 and 2.52 for N-methylephedrine, 4.79 and 6.48 for terbutaline, and 0 and 1.74 for hexobarbital. Analysis of the cyclodextrins showed differences in the distribution of components and in the UV spectra.

Five commercially available dimethyl β-cyclodextrins were evaluated by measuring the reproducibility in enantioselectivity for the enantiomers of thirteen compounds [82]. The compounds included several important pharmaceutical agents such as terbutaline, salbutamol, atenolol, and bupivacaine. The reproducibility of the enantioselectivity was found to vary between the compounds. In some cases only very small differences in enantioselectivity were seen whereas in other cases the differences were significant. The composition of the cyclodextrin samples was investigated by separating the components using HPLC and measuring them with refractive index detection. The HPLC analysis showed two main components which were present at different levels in the different cyclodextrins. For one of the cyclodextrin samples these two components were separated by preparative chromatography. The two preparative HPLC fractions were analysed using both proton and ^{13}C NMR, and time of flight MS. The data indicated the first component to be the heptakis(2,6-di-O-methyl)-β-cyclodextrin and the second component was believed to be hexakis(2,6-di-O-methyl)-mono(2,3,6-tri-O-methyl)-β-cyclodextrin.

5.9 Conclusion

Cyclodextrins are popular for the separation of enantiomers by CE because the range of enantioselectivities available and their ease of use. Cyclodextrins have been successfully used with a wide range of analyte classes and analytical applications.

References

[1] Gübitz, G.; Schmid, M.G. Chiral separation principles in capillary electrophoresis, *J. Chromatogr. A.* **1997**, *792*, 179–225.

[2] Szejtli, J. Introduction and General Overview of Cyclodextrin Chemistry, *Chem. Rev.* **1998**, *98*, 1743–1753.

[3] Szejtli, J. Cyclodextrin Technology, Kluwer academic publishers, Dordrecht, **1988**.

[4] Loftsson, T.; Brewster, M.E. Cyclodextrins as Pharmaceutical Excipients, *Pharm. Tech. Europe.* **1997**, *5*, 26–34.

[5] Wacker Chemie – private communication.

[6] Taghvaei, M.; Stewart, G.H. β-Cyclodextrin Solubility in Reversed-Phase High-Performance Liquid Chromatographic Eluents, *Anal. Chem.* **1991**, *63*, 1902–1904.

[7] Chatjigakis, A.K.; Donzé, C.; Coleman, A.W.; Cardot, P. Solubility Behaviour of β-Cyclodextrin in Water/Cosolvent Mixtures, *Anal. Chem.* **1992**, *64*, 1632–1634.

[8] Pharr, D.Y.; Fu, Z.S.; Smith, T.K.; Hinze, W.L. Solubilization of Cyclodextrins for analytical applications, *Anal. Chem.* **1989**, *61*, 275–279.

[9] Yusuff, N.; York, P. Spironolactone-cyclodextrin complexes: phase solubility and ultrafiltration studies, *Int. J. Pharm.* **1991**, *73*, 9–15.

[10] Uekama, K.; Narisawa, S.; Hirayama, F.; Otagiri, M. Improvement of dissolution and absorption characteristics of benzodiazepines by cyclodextrin complexation, *Int. J. Pharm.* **1983**, *16*, 327–338.

[11] Inoue, Y.; Hakushi, T.; Liu, Y.; Tong, L.-H.; Shen, B.-J.; Jin, D.-S. Thermodynamics of Molecular Recognition by Cyclodextrins. 1. Calorimetric Titration of Inclusion Complexation of Naphthalenesulfonates with α-, β-, and γ-Cyclodextrins: Enthalpy-Entropy Compensation, *J. Am. Chem. Soc.* **1993**, *115*, 475–481.

[12] Uekama, K.; Fujinaga, T.; Hirayama, F.; Otagiri, M.; Yamasaki, M. Inclusion complexations of steroid hormones with cyclodextrins in water and in solid phase, *Int. J. Pharm.* **1982**, *10*, 1–15.

[13] Czugler, M.; Eckle, E.; Stezowski, J. Crystal and Molecular Structure of a 2,6-Tetradeca-O-methyl-β-cyclodextrin – Adamantol 1:1 Inclusion Complex. *J.C.S. Chem. Comm.* **1981**, 1291–1292.

[14] Greatbanks, D.; Pickford, R. Cyclodextrins as Chiral Complexing Agents in Water, and their Application to Optical Purity Measurements. *Magn. Reson. in Chem.* **1987**, *25*, 208–215.

[15] Li, S.; Purdy, W.C. Circular Dichoism, Ultraviolet, and Proton Nuclear Magnetic Resonance Spectroscopic Studies of the Chiral Recognition Mechanism of β-cy-

clodextrin, *Anal. Chem.* **1992**, *64*, 1405–1412.

[16] Chankvetadze, B.; Schulte, G.; Bergenthal, D.; Blaschke, G. Comparative capillary electrophoresis and NMR studies of enantioseparation of dimethindene with cyclodextrins, *J. Chromatogr. A.* **1998**, *798*, 315–323.

[17] Valkó, I.E.; Billiet, H.A.H.; Frank, J.; Luyben, K.Ch.A.M. Factors Affecting the Separation of Mandelic Acid Enantiomers by Capillary Electrophoresis, *Chromatographia* **1994**, *38*, 730–736.

[18] Ferguson, P.D.; Goodall, D.M.; Loran, J.S Systematic approach to the treatment of enantiomeric separations in capillary electrophoresis and liquid chromatography III. A binding constant-retention factor relationship and effects of acetonitrile on the chiral separation of tioconazole, *J. Chromatogr. A* **1996**, *745*, 25–35.

[19] Gratz, S.R.; Stalcup, A.M. Enantiomeric Separations of Terbutaline by CE with a Sulfated β-Cyclodextrin Chiral Selector: A Quantitative Binding Study. *Anal. Chem.* **1998**, *70*, 5166–5171.

[20] Reijenga, J.C.; Ingelse, B.A.; Everaerts, F.M. Thermodynamics of chiral selectivity in capillary electrophoresis: separation of ibuprofen enantiomers with β-cyclodextrin, *J. Chromatogr. A* **1997**, *792*, 371–378.

[21] Ma, S.; Horváth, C. Capillary zone electrophoresis at subzero temperature II: Chiral separation of biogenic amines, *Electrophoresis* **1997**, *18*, 873–883.

[22] Rawjee, Y.Y.; Staerk, D.U.; Vigh, G. Capillary electrophoretic chiral separations with cyclodextrin additives. I. Acids: chiral selectivity as a function of pH and the concentration of β-cyclodextrin for fenoprofen and ibuprofen, *J. Chromatogr* **1993**, *635*, 291–306.

[23] Beesley, T.E.; Scott, R.P.W. Chiral Chromatography, J. Wiley and Sons, New York, **1998**.

[24] Zuowski, J.; Sybilska, D.; Bojarski, J. Applications of α- and β-cyclodextrin and heptakis(2,6-di-O-methyl)-β-cyclodextrin as mobile phase components for the separation of some chiral barbiturates into enantiomers by reversed-phase high-performance liquid chromatography, *J. Chromatogr.* **1986**, *364*, 225–232.

[25] Horváth, Cs.; Melander, W.; Nahum, A. Measurement of association constants for complexes by reversed-phase high-performance liquid chromatography, *J. Chromatogr.* **1979**, *186*, 371–403.

[26] Mayer, S.; Schleimer, M.; Schurig, V. Dual Chiral Recognition System Involving Cyclodextrin Derivatives in Capillary Electrophoresis, *J. Microcol. Sep.* **1994**, *6*, 43–48.

[27] Schmitt, T.; Engelhardt, H. Derivatised Cyclodextrins for the Separation of Enantiomers in Capillary Electrophoresis, *J. High Resolut. Chromatogr.* **1993**, *16*, 525–529.

[28] Yoshinaga, M.; Tanaka, M. Use of selectively methylated β-cyclodextrin derivatives in chiral separation of dansylamino acids by capillary zone electrophoresis, *J. Chromatogr. A* **1994**, *679*, 359–365.

[29] Aturki, Z.; Desiderio, C.; Mannina, L.; Fanali, S. Chiral separations by capillary zone electrophoresis with the use of cya-

noethylated-β-cyclodextrin as chiral selector, *J. Chromatogr. A* **1998**, *817*, 91–104.

[30] Chankvetadze, B.; Schulte, G.; Blashke, G. Reversal of enantiomer order in capillary electrophoresis using charged and neutral cyclodextrins, *J. Chromatogr. A* **1996**, *732*, 183–187.

[31] Andrisano, V.; Gotti, R.; Cavrini, V.; Tumiatti, V.; Felix, G.; Wainer, I.W. Capillary electrophoretic and high-performance liquid chromatographic studies of the enantioselective separation of α_1-adrenoreceptor antagonists, *J. Chromatogr. A* **1998**, *803*, 189–195.

[32] Bechet, I.; Paques, P.; Fillet, M.; Hubert, P.; Crommen, J. Chiral separation of basic drugs by capillary zone electrophoresis with cyclodextrins additives, *Electrophoresis* **1994**, *15*, 818–823.

[33] A. Guttman, S. Brunet, N. Cooke, Capillary Electrophoresis Separation of Enantiomers Using Cyclodextrin Array Chiral Analysis., *LC-GC Intl.* February **1996**, 88–100.

[34] Liu, L.; Nussbaum, M.A. Systematic screening approach for chiral separations of basic compounds by capillary electrophoresis with modified cyclodextrins, *J. Pharm. Biomed. Anal.* **1999**, *19*, 679–694.

[35] lin, B.; Zhu, X.; Koppenhoefer, B.; Epperlein, U. Investigation of 123 Chiral Drugs by Cyclodextrin-Modified Capillary Electrophoresis, *LC-GC Intl.* January **1997**, 40–46.

[36] Wren, S.A.C. Chiral separation in capillary electrophoresis, *Electrophoresis* **1995**, *16*, 2127–2131.

[37] Schmitt, T.; Engelhardt, H. Charged and Uncharged Cyclodextrins as Chiral Selectors in Capillary Electrophoresis, *Chromatographia* **1993**, *37*, 475–481.

[38] Juvancz, Z.; Jicsinsky, L.; Markides, K.E. Phosphated Cyclodextrins as New Acidic Chiral Additives for Capillary Electrophoresis., *J. Microcol. Sep.* **1997**, *9*, 581–589.

[39] Juvancz, Z.; Markides, K.E.; Jicsinsky, L. Chiral Analysis of Metoprolol and Its By-Products by Capillary Electrophoresis, *J. Microcol. Sep.* **1999**, *11*, 716–722.

[40] Tait, R.J.; Thompson, D.O.; Stella, V.J.; Stobaugh, J.F. Sulfobutyl Ether β-Cyclodextrin as a Chiral Discriminator for Use with Capillary Electrophoresis, *Anal. Chem.* **1994**, *66*, 4013–4018.

[41] Desiderio, C.; Fanali, S. Use of negatively charged sulfobutyl ether-β-cyclodextrin for enantiomeric separation by capillary electrophoresis, *J. Chromatogr. A* **1995**, *716*, 183–196.

[42] Vargas, M.G.; Vander Heyden, Y.; Maftouh, M.; Massart, D.L. Rapid development of the enantiomeric separation of β-blockers by capillary electrophoresis using an experimental design approach, *J. Chromatogr. A* **1999**, *855*, 681–693.

[43] Rickard, E.C.; Bopp, R.J.; Skanchy, D.J.; Chetwyn, K.L.; Pahlen, B.; Stobaugh, J.F. Role of Capillary Electrophoresis Methods in the Drug Development Process, *Chirality* **1996**, *8*, 108–121.

[44] Francotte, E.; Brandel, L.; Jung, M. Influence of the degree of substitution of cyclodextrin sulphobutyl ether derivatives on enantioselective separations by electrokinetic chromatography, *J. Chromatogr. A.* **1997**, *792*, 379–384.

[45] Stalcup, A.M.; Gahm, K.H. Application of Sulfated Cyclodextrins to Chiral Separations by Capillary Zone Electrophoresis, *Anal. Chem.* **1996**, *68*, 1360–1368.

[46] Vincent, J.B.; Sokolowski, A.D.; Nguyen, T.V.; Vigh, G. A Family of Single-Isomer Chiral Resolving Agents for Capillary Electrophoresis. 1. Heptakis (2,3-diacetyl-6-sulfato)-β-cyclodextrin, *Anal. Chem.* **1997**, *69*, 4226–4233.

[47] Vincent, J.B.; Kirby, D.M.; Nguyen, T.V.; Vigh, G. A Family of Single-Isomer Chiral Resolving Agents for Capillary Electrophoresis. 2. Hepta-6-sulfato-β-cyclodextrin, *Anal. Chem.* **1997**, *69*, 4419–4428.

[48] Cai, H.; Nguyen, T.V.; Vigh, G. A Family of Single-Isomer Chiral Resolving Agents for Capillary Electrophoresis. 3. Heptakis (2,3-dimethyl-6-sulfato)-β-cyclodextrin, *Anal. Chem.* **1998**, *70*, 580–589.

[49] Zhu, W.; Vigh, G. Capillary Electrophoretic Separation of the Enantiomers of Weak Acids in a High pH Background Electrolyte Using the New, Single-Isomer, Octakis(2,3-diacetyl-6-sulfato)-γ-cyclodextrin as Chiral Resolving Agent, *J. Microcol. Sep.* **2000**, *12*, 167–171.

[50] Nardi, A.; Eliseev, A.; Bocek, P.; Fanali, S. Use of charged and neutral cyclodextrins in capillary zone electrophoresis: enantiomeric resolution of some 2-hydroxy acids, *J. Chromatogr.* **1993**, *638*, 247–253.

[51] Haynes, J.L.; Shami, S.A.; O'Keefe, F.; Darcey, R.; Warner, I.M. Cyclodextrin derivative for chiral separations, *J. Chromatogr. A* **1998**, *803*, 261–271.

[52] Schulte, G.; Chankvetadze, B.; Blaschke, G. Enantioseparation in capillary electrophoresis using 2-hydroxypropyltrimethylammonium salt of β-cyclodextrin as a chiral selector, *J. Chromatogr. A* **1997**, *771*, 259–266.

[53] Jakubetz, H.; Juza, M.; Schurig, V. Electrokinetic chromatography employing an anionic and a cationic β-cyclodextrin derivative, *Electrophoresis* **1997**, *18*, 897–904.

[54] Bunke, A.; Jira, T. Use of cationic cyclodextrin for enantioseparation by capillary electrophoresis, *J. Chromatogr. A* **1998**, *798*, 275–280.

[55] Galaverna, G.; Corradini, R.; Dossena, A.; Marchelli, R.; Vecchio, G. Histamine-modified β-cyclodextrins for the enantiomeric separation of dansyl-amino acids in capillary electrophoresis, *Electrophoresis* **1997**, *18*, 905–911.

[56] Lelièvre, F.; Gueit, C.; Gareil, P.; Bahaddi, Y.; Galons, H. Use of zwitterionic cyclodextrin as a chiral agent for the separation of enantiomers by capillary electrophoresis, *Electrophoresis* **1997**, *18*, 891–896.

[57] Nishi, H.; Fukuyama, T.; Terabe, S. Chiral separation by cyclodextrin-modified micellar electrokinetic chromatography, *J. Chromatogr.* **1991**, *553*, 503–516.

[58] Schmid, M.G.; Wirnsberger, K.; Jira, T.; Bunke, A.; Gübitz, G. Capillary Electrophoretic Chiral Resolution of Vicinal Diols by Complexation With Borate and Cyclodextrin: Comparative Studies on Different Cyclodextrin Derivatives, *Chirality* **1997**, *9*, 153–156.

[59] Billiot, E.; Wang, J.; Warner, I.M. Improved chiral separation using achiral

modifiers in cyclodextrin modified capillary zone electrophoresis, *J. Chromatogr. A* **1997**, *773*, 321–329.

[60] Okafo, G.N.; Bintz, C.; Clark, S.E.; Camilleri, P. Micellar Electrokinetic Chromatography in a Mixture of Taurodeoxycholic Acid and β-Cyclodextrin, *J. Chem. Soc. Chem. Com.* **1992**, 1189–1192.

[61] Cooper, A.; Nutley, M.A.; Camilleri, P. Microcalorimetry of Chiral Surfactant-Cyclodextrin Interactions, *Anal. Chem.* **1998**, *70*, 5024–5028.

[62] Zhou, L.; Trubig, J.; Dovletoglu, A.; Locke, D.C. Enantiomeric separation of the novel growth hormone secretagogue MK-0677 by capillary zone electrophoresis, *J. Chromatogr. A* **1997**, *773*, 311–320.

[63] Jira, T.; Bunke, A.; Karbaum, A. Use of chiral and achiral ion-pairing reagents in combination with cyclodextrins in capillary electrophoresis, *J. Chromatogr. A* **1998**, *798*, 281–288.

[64] Fillet, M.; Hubert, P.; Crommen, J. Enantioseparation of nonsteroidal anti-inflammatory drugs by capillary electrophoresis using mixtures of anionic and uncharged β-cyclodextrins as chiral additives, *Electrophoresis* **1997**, *18*, 1013–1018.

[65] Fillet, M.; Fotsing, L.; Crommen, J. Enantioseparation of uncharged compounds by capillary electrophoresis using mixtures of anionic and neutral β-cyclodextrin derivatives, *J. Chromatogr. A* **1998**, *817*, 113–119.

[66] Rudaz, S.; Veuthey, J.-L.; Desiderio, C.; Fanali, S. Enantioseparation of Venlafaxine and O-Desmethylvenlafaxine by Capillary Electrophoresis with Mixed Cyclodextrins, *Chromatographia* **1999**, *50*, 369–372.

[67] Rudaz, S.; Cherkaoui, S.; Dayer, P.; Fanali, S.; Veuthey, J.-L. Simultaneous stereoselective analysis of tramadol and its main phase I metabolites by on-line capillary zone electrophoresis-electrospray ionization mass spectrometry. *J. Chromatogr. A* **2000**, *868*, 295–303.

[68] Lanz, M.; Brenneisen, R.; Thormann, W. Enantioselective determination of 3,4-methylenedioxymethamphetamine and two of its metabolites in human urine by cyclodextrin-modified capillary zone electrophoresis, *Electrophoresis* **1997**, *18*, 1035–1043.

[69] Silverman, C. Chiral separations by capillary electrophoresis in process chemistry, *J. Cap. Elec.* **1997**, *4*, 181–187.

[70] Lurie, I.S.; Klein, R.F.X.; Dal Cason, T.A.; LeBelle, M.J.; Brenneisen, R.; Weinberger, R.E. Chiral Resolution of Cationic Drugs of Forensic Interest by Capillary Electrophoresis with Mixtures of Neutral and Anionic Cyclodextrins, *Anal. Chem.* **1994**, *66*, 4019–4026.

[71] Greenaway, M.; Okafo, G.N.; Camilleri, P.; Dhanak, D. A Sensitive and Selective Method for the Analysis of Complex mixtures of Sugars and Linear Oligosaccharides, *J. Chem. Soc. Chem. Commun.* **1994**, 1691–1692.

[72] Skanchy, D.J.; Wilson, R.; Poh, T.; Xie, G.-H.; Demarest, C.W.; Stobaugh, J.F. Resolution of acylated dipeptide stereoisomers by capillary electrophoresis using sulfobutylether derivatized β-cyclodextrin, *Electrophoresis* **1997**, *18*, 985–995.

[73] Verleysen, K.; Sabah, S.; Scriba, G.; Chen, A.; Sandra, P. Evaluation of the enantioselective possibilities of sulfated cyclodextrins for the separation of aspartyl di- and tripeptides in capillary electrophoresis, *J. Chromatogr. A* **1998**, *824*, 91–97.

[74] Fanali, S.; Flieger, M.; Steinerova, N.; Nardi, A. Use of cyclodextrins for the enantioselective separation of ergot alkaloids by capillary zone electrophoresis, *Electrophoresis* **1992**, *13*, 39–43.

[75] Tahara, S.; Okayama, A.; Kitada, Y.; Watanabe, T.; Nakazawa, H.; Kakehi, K.; Hisamatu, Y. Enantiomeric separation of atropine in *Scopolia* extract and *Scopolia Rhizome* by capillary electrophoresis using cyclodextrins as chiral selectors, *J. Chromatogr. A* **1999**, *848*, 465–471.

[76] Mateus, L.; Cherkaoui, S.; Christen, P.; Veuthey, J.-L. Enantioseparation of atropine by capillary electrophoresis using sulfated β-cyclodextrin: application to a plant extract, *J. Chromatogr. A* **2000**, *868*, 285–294.

[77] Garrison, A.W.; Schmitt, P.; Kettrup, A. Separation of phenoxy herbicides and their enantiomers by high-performance capillary electrophoresis, *J. Chromatogr. A* **1994**, *688*, 317–327.

[78] Hadley, M.R.; Gabriac, S.D.; Hutt, A.J. Resolution of Enantiomeric N-Oxides by Capillary Electrophoresis using Cyclodextrins as Chiral Selectors, *Chirality* **1999**, *11*, 409–415.

[79] Valenzuela, F.A.; Green, T.K.; Dahl, D.B. Enantiomeric Separation of Sulfonium Ions by Capillary Electrophoresis Using Neutral and Charged Cyclodextrins, *Anal. Chem.* **1998**, *70*, 3612–3618.

[80] Crego, A.L.; García, M.A.; Marina, M.L. Enantiomeric Separation of Chiral Polychlorinated Biphenyls by Micellar Electrokinetic Chromatography Using Mixtures of Bile Salts and Sodium Dodecyl Sulphate with and without γ-Cyclodextrin in the Separation Buffer, *J. Microcol. Sep.* **2000**, *12*, 33–40.

[81] Szemán, J.; Roos, N.; Csabai, K. Ruggedness of enantiomeric separation by capillary electrophoresis and high-performance liquid chromatography with methylated cyclodextrins as chiral selectors, *J. Chromatogr. A* **1997**, *763*, 139–147.

[82] Otsuka, K.; Honda, S.; Kato, J.; Terabe, S.; Kimata, K.; Tanaka, N. Effects of compositions of dimethyl-β-cyclodextrins on enantiomer separations by cyclodextrin modified capillary zone electrophoresis, *J. Pharm. Biomed. Anal.* **1998**, *17*, 1177–1190.

Other Chiral Selectors

6.1 Introduction

Chapter 5 covered the applications of cyclodextrins for the separation of enantiomers in CE and some of the benefits. Whilst the cyclodextrins easily constitute the most widely used group of selectors for the separation of enantiomers in CE, a range of other selectors have also been employed. Some of these other selectors have been reported to offer significant benefits over cyclodextrins for example in terms of greater enantioselectivity. As is the case with the cyclodextrins many of the other chiral selectors used in CE are also naturally occurring species or are derived from them. Chapter 6 covers some of the other selectors which have been employed for the separation of enantiomers along with some examples of their application.

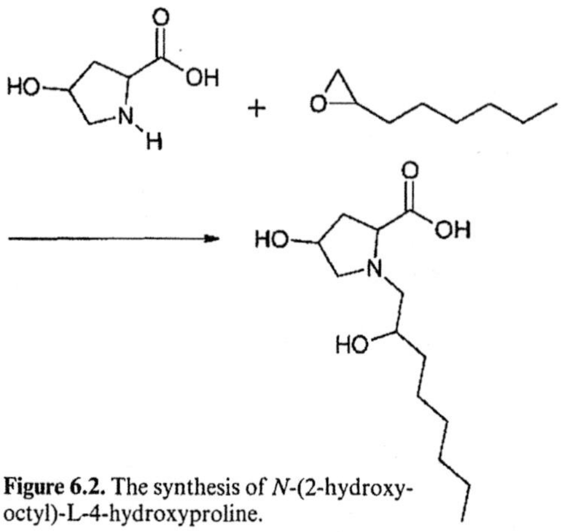

6.2 Ligand Exchange Selectors

The use of ligand exchange procedures was one of the first approaches reported for the separation of enantiomers in CE. For example the enantiomers of some dansylated amino acids were resolved by the use of the copper (II) complexes of either L-histidine or aspartame in 1985 and 1987 respectively [1, 2]. The dansylated amino acids are thought to exchange with the L-histidine to form diastereoisomeric copper (II) species with different stabilities. The dansylated amino acid enantiomers separate because the electrophoretic mobilities of the free and complexed forms are different. Replacing L-histidine in the copper (II) complex with D-histidine lead to a reversal in the migra-

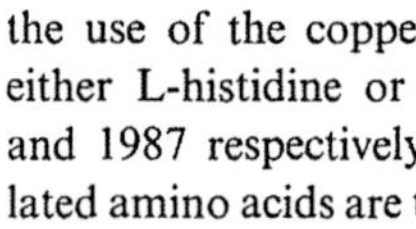

Figure 6.1. Phenylalanine.

Figure 6.2. The synthesis of *N*-(2-hydroxy-octyl)-L-4-hydroxyproline.

tion order of the dansylated amino acid enantiomers.

Following this pioneering work there have been relatively few examples of the use of ligand exchange for enantiomer separation in CE. More recently however there have been investigations into the separation of some of the enantiomers of underivatised amino acids. Examination of the native amino acids requires the use of UV absorbance detection in place of laser induced fluorescence. This change of detection system simplifies analytical procedures but at the cost of a large decrease in sensitivity. For example the resolution between the enantiomers of eleven amino acids produced using the copper (II) complexes of L-proline and L-4-hydroxyproline has been measured [3]. The copper (II) complexes were produced using 2 moles of the amino acid ligand per mole of copper (II) and better results were obtained using L-4-hydroxyproline instead of L-proline. With DL-phenylalanine (Figure 6.1) the enantioselectivity increased with increasing concentration of copper (II) L-4-hydroxyproline, with a plateau being reached at a concentration of around 80 mM L-4-hydroxyproline. The resolution also varied with pH of the buffer system with the optimum value being around pH 4. Examination of the

amino acids under the optimum conditions for phenylalanine gave resolution values ranging from 0 with α-phenylglycine, to 2.76 with *p*-tyrosine. The D-enantiomers of the amino acids always migrated faster than the L-enantiomers demonstrating that the D-aminoacids form the stronger copper (II) complexes. The addition of SDS to the running buffer lead to a reversal in the migration order of the amino acid enantiomers. The selectivity and resolution pass through zero as the SDS concentration is increased and the migration order of the amino acid enantiomers swaps around.

The ligand exchange work with the copper (II) complexes of L-4-hydroxyproline was extended by the preparation and examination of *N*-(2-hydroxy-octyl)-L-4-hydroxyproline (HO-L-Hypro) as a ligand [4]. The HO-L-Hypro ligand was prepared by the reaction of L-4-hydroxyproline with 1,2-Epoxyoctane, Figure 6.2, and was examined on the assumption that the long hydrophobic side chain would improve resolution.

The HO-L-Hypro ligand was found to give significantly better resolution than L-4-hyroxyproline for a range of amino acid analytes, and at a much lower concentration (by one order of magnitude). For example the resolution between the enantio-

Original

0009-5893/00/02   78-16 $ 03.00/0

mers of phenylalanine was 1.34 as opposed to 1.19, and the resolution between the enantiomers of dihydroxyphenylalanine (DOPA) was 2.70 as opposed to 1.19. The higher resolution meant that it was possible to measure levels of D-DOPA down to 0.03% in the presence of L-DOPA. The buffer pH is an important parameter to optimise as it controls the degree of dissociation of both the exchange ligand and the analyte according to their respective pI values. For example the resolution of histidine enantiomers was optimised at about pH 6, whilst the resolution of α-methyldopa enantiomers at about pH 4.3.

The combination of SDS and the copper (II) complex of L-hydroxyproline has also been employed for the separation of the positional and optical isomers of some tryptophan derivatives [5]. The copper (II) complex of L-hydroxyproline and SDS (or other anionic surfactants) and were also used to separate the enantiomers of the o-, m-, and p-positional isomers of fluorophenyalanine [6]. Again the migration orders of the enantiomers could be reversed by altering the concentration of the anionic surfactant.

6.3 Chiral Surfactants

A wide range of chiral surfactants have been employed for the separation of enantiomers by CE and some of them will be discussed in this section. The surfactants employed are both naturally occurring and synthetic, and have been used for many different analytes. The analyte enantiomers are thought to separate via partitioning into micelles formed by chiral surfactants. Enantioseparation requires both that individual enantiomers have different affinities for the micelle, and that the micelle has a different electrophoretic mobility to that of the free enantiomers.

6.3.1 Synthetic Chiral Surfactants

The use of synthetic chiral surfactants with long alkyl chains followed soon after the introduction of Micellar ElectroKinetic Chromatography (MEKC) using SDS for the separation of diastereoisomers. Synthetic chiral surfactants allow for the examination of a range of potential chiral selectors and mean that the stereochemistry of the selector itself can also be altered. The ability to employ both R and S chiral selectors gives the possibility for

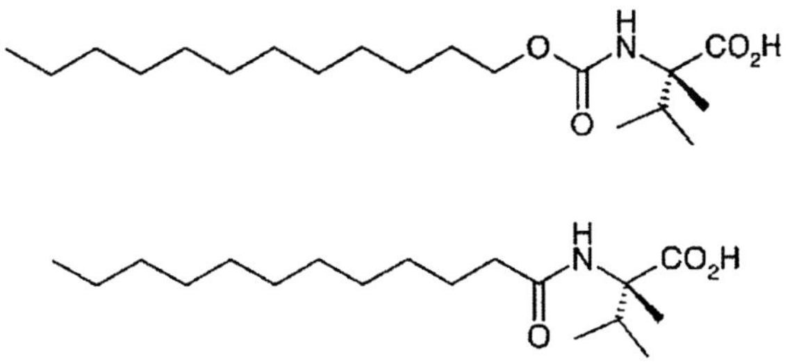

Figure 6.3. *(S)* N-dodecoxycarbonylvaline and *(S)* N-dodecanoylvaline.

the analyst to choose the migration order of the analyte enantiomers.

An early example of synthetic chiral selectors in CE is sodium N-dodecanoyl-L-valinate (SDVal). SDVal is prepared by the reaction of valine with dodecanoyl chloride, and has been employed for the separation of phenylthiohydantoin (PTH) derivatives of some amino acids. For example a 25 mM solution of SDVal in phosphate buffer at pH 7 was used to partially resolve the PTH derivatives of norvaline, tryptophan, and norleucine [7]. The addition of 5 M urea to the buffer is found to be beneficial by significantly reducing the degree of peak tailing. The resolution between these and other PTH derivatised amino acids is improved by the use of a co-micellar system containing SDS and SDVal [8]. The same SDS and SDVal co-micellar system was also used for the separation of the enantiomers of benzoin and warfarin.

Chiral selectors which are closely related to SDVal have also been synthesised and evaluated. For example the performance of the (R) and (S) enantiomers of N-dodecoxycarbonylvaline was compared to that of SDVal (Figure 6.3) [9].

The small structural change between these two chiral surfactants was designed to reduce the background UV absorbance, but was also found to give rise to significant differences in the enantioselectivity. The two chiral surfactants were used to analyse twelve pharmaceutical compounds containing amino groups. *(S)* N-dodecoxycarbonylvaline was found to give higher enantioselectivity than *(S)* N-dodecanoylvaline for ten of the twelve pharmaceuticals. The benefits of having both enantiomers of N-dodecoxycarbonylvaline were demonstrated by showing that a change from the (R) to the (S) form of the surfactant gave a reversal in the migration order of the benzoin enantiomers. In another experiment the enantioselectivity for the enantiomers of N-methylpseudoephedrine was measured using a range of buffers containing different propor-

tions of (R) and *(S)* N-dodecoxycarbonylvaline. The enantioselectivity for N-methylpseudoephedrine was found to vary linearly with the percentage of the (R) surfactant used in the separation buffer. There was no enantioselectivity and a change over in migration order when a mixture of 50% (R) and 50% *(S)* surfactant was used. The *(S)* N-dodecoxycarbonylvaline surfactant was successfully applied to the separation of ephedrine in urine samples. Changing the buffer pH, and so the analyte charge, was found to alter both the capacity factor and the enantioselectivity.

Unsaturated derivatives of amino acids have been used to prepare polymeric chiral surfactants. For example sodium N-undecylenyl-L-valine (L-SUV) was used to synthesise poly(sodium N-undecylenyl-L-valinate) (poly-L-SUV) [10, 11]. One advantage of the polymeric surfactants over their monomeric counterparts is that they can be used at low concentrations as their critical micellar concentration is practically zero. The chiral surfactant poly-L-SUV was used to resolve the enantiomers of a number of analytes including 1,1'-binaphthyl-2,2'-diol and the dinitrobenzoyl derivatives of some amino acids. The polymeric surfactant produced by the D form of the monomer, poly-D-SUV, was also used in a mixed chiral selector system by combination with γ-cyclodextrin. The mixed chiral selector system gave better resolution of the enantiomers of 1,1'-binaphthyl-2,2'-diol than could be achieved by the use of either of the chiral selectors in isolation. Replacement of poly-D-SUV with poly-L-SUV in the mixed surfactant/cyclodextrin selector system lead to a loss of resolution.

6.3.2 Naturally Occurring Chiral Surfactants

Several naturally occurring surfactants have been successfully employed for the separation of enantiomers. Amongst the most popular of these natural surfactants

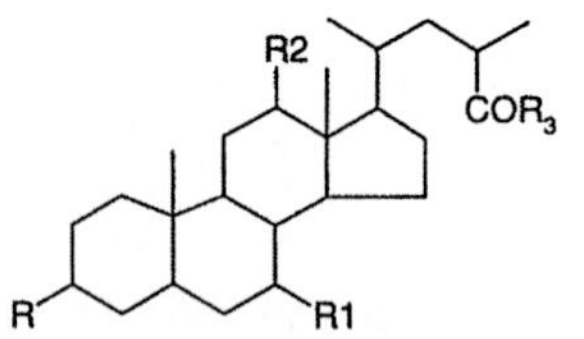

Figure 6.4. The bile salts sodium cholate, sodium taurocholate, sodium deoxycholate, and sodium taurodeoxycholate.

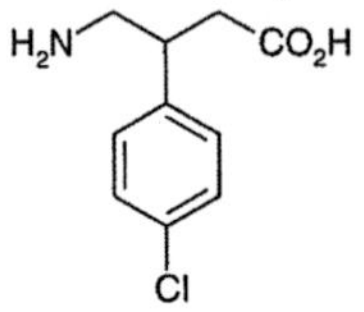

Figure 6.5. Baclofen.

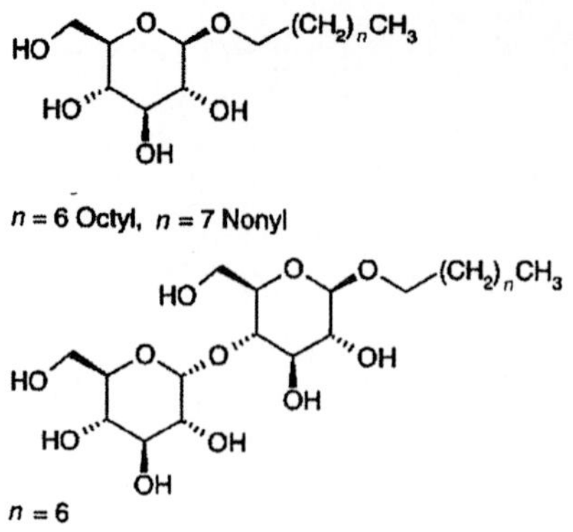

n = 6 Octyl, *n* = 7 Nonyl

n = 6

Figure 6.6. The alkylglcoside surfactants *n*-octyl-β-D-glucopyranoside, *n*-nonyl-β-D-glucopyranoside, and *n*-octyl-β-D-maltopyranoside.

are the bile acid salts sodium cholate, taurocholate, deoxycholate and taurodeoxycholate. The bile acids have a hydroxy substituted steroid backbone and are shown in Figure 6.4.

The bile acids were used at a concentration of 50 mM to resolve the enantiomers of a range of analytes including diltiazem, trimetoquinol, 1-naphthylethylamine, and some carboline derivatives [12]. The resolution obtained varied with the analyte and bile acid employed but none of the bile salts proved ideal for all of the analytes. The resolution also depended upon the pH of the buffer (and so the charge on the analyte).

Sodium taurodeoxycholate (STDC) was used to resolve the enantiomers of the local anaesthetics mepivacaine, bupivacaine, and prilocaine [13]. The resolution between the enantiomers was found to depend upon the concentration of STDC with different optimum concentrations being found for different analytes. The optimum concentration for mepivacaine was 30 mM and that for prilocaine 45 mM. The differences in the optimum concentrations are thought to relate to differences in the hydrophobicities of the analytes, and a complexation model similar to that used for cyclodextrins has been proposed (see Chapter 3). The migration order of the individual analyte enantiomers was established by spiking experiments, with the (*S*) enantiomers found to have a stronger affinity for the micelles than the (*R*) enantiomers. The enantiomers of bupivacaine could be resolved but the low concentrations of STDC required gave long analysis times. The affinity of the local anaesthetics for the micelles could be manipulated by the use of co-micellar systems containing both STDC and

the non-ionic surfactant Brij-35 (polyoxethylene(23)-dodecanol).

Bile acids have also been employed in mixed chiral selector systems. For example a system employing both sodium deoxycholate (SDC) and β-cyclodextrin was used to separate the enantiomers of some derivatised amino acids [14]. The dansyl and 1-cyano-2-substituted-benz[L]isoindole (CBI) derivatives of a range of amino acids were separated using 50 mM of SDC and 20 mM of β-cyclodextrin at pH 7. The resolution between the CBI derivatives of histidine and glutamic acid was better with SDC/cyclodextrin than could be achieved with mixed SDS/cyclodextrin selector systems. The SDC/β-cyclodextrin system was also used to separate the enantiomers of the CBI derivatives of baclofen (Figure 6.5) and some of its amino phosphate analogues. The two derivatisation approaches were applied to the amino acids produced from the hydrolysis of D-Phe[7] bradykinin (Arg-Pro-Pro-Gly-Phe-Ser-D-Phe-Phe-Arg).

Another natural surfactant which has been used for the separation of enantiomers is digitonin. Digitonin is a neutral surfactant which has five sugar units linked to a steroid backbone. Digitonin was used in combination with SDS to resolve the phenylthiohydantoin derivatives of the DL amino acids tryptophan, norleucine, norvaline, valine, α-aminobutyric acid, and alanine [15].

Other sugar containing surfactants have also been employed for the separation of enantiomers by CE. For example the alkylglycoside surfactants *n*-octyl-β-D-glucopyranoside, *n*-nonyl-β-D-glucopyranoside, and *n*-octyl-β-D-maltopyranoside (Figure 6.6) were evaluated for the separation of the dansylated derivatives of

some DL amino acids, binaphthyl phosphate, warfarin, and bupivacaine [16].

The separations were carried out at pH 6.5 using sodium phosphate buffers at concentrations of 150 mM or higher and using surfactant concentrations in the range 10 to 80 mM. Of the amino acids examined phenylalanine, leucine, methionine and valine could be separated by all three of the surfactants. Tryptophan could only be separated by octylglucopyranoside. And aspartic acid, glutamic acid, serine and threonine were unresolved by any of the three surfactants over a wide range of concentrations. With the amino acid enantiomers which could be separated by the alkylglycosides, the degree of resolution varied with the surfactant concentration. With several of the amino acids there was an optimum surfactant concentration and this concentration can be related to the hydrophobicity of the amino acid. The hydrophobic amino acids had low optimum surfactant concentrations and the more hydrophilic amino acids had higher optimum surfactant concentrations. The enantiomers of 1,1′-binaphthyl-2,2′-diylhydrogen phosphate were resolved by all three alkylglycoside surfactants with the octylmaltopyranoside being effective even below its critical micellar concentration. The enantiomers of warfarin could only be satisfactorily resolved by using 80 mM of the octylmaltopyranoside and increasing the sodium phosphate buffer concentration to 250 mM. The enantiomers of bupivacaine could only be resolved using 150 mM of the octylmaltopyranoside and again the resolution was improved by increasing the concentration of sodium phosphate from 50 to 150 and then to 200 mM.

6.4 Alkaloids

The alkaloids are another class of naturally occurring compounds which have been employed for the separation of enantiomers using capillary electrophoresis. For example several ergot alkaloids were examined as potential chiral selectors for the separation of the enantiomers of some aromatic hydroxy acids [17]. The ergot alkaloids are structurally related to (5*R*)-lysergic acid and have the potential to interact with aromatic hydroxy acids via a range of mechanisms including electrostatic and π-π interactions. Eight of the ergot alkaloids were screened for the resolution of the enantiomers of mandelic acid, *p*-hydroxymandelic acid, 3,4-dihydroxy-mandelic acid, vanilmandelic acid, and tropic acid. The best results were found using the 1-allyl derivative of (5*R*, 8*S*, 10*R*)-terguride (Figure 6.7).

The ergot alkaloids absorb strongly in the UV spectrum and so the experimental conditions have to be chosen carefully to avoid interference with the analytes. A neutrally coated capillary was used and the pH chosen to ensure that the chiral selector moved in the opposite direction to that of the analytes (i.e. away from the detector). The capillary was rinsed with buffer containing the ergot alkaloids but the separation vials only contained the other buffer components. The separation was pH dependent and it was concluded that the allyl terguride only showed enantioselectivity for the dissociated forms of the acid analytes. Using a pH of 4.2 the enantiomers of all the acids except tropic acid were baseline resolved with a 25 mM concentration of 1-allylterguride. The equilibrium constants for the formation of the acid/1-allylterguride complexes were determined by measuring the apparent electrophoretic mobilities at a range of concentrations. It was also shown that for some acids the enantioselectivity (as measured by the ratio of the equilibrium constants) could be significantly changed by switching to a pH 5.3 buffer system containing 50% methanol. For example the enantioselectivity for tropic acid increased from 1.01 to 1.07 in the methanol containing buffer leading to baseline resolution.

In an interesting twist the relationship between the analyte and the chiral selector was switched and the enantiomers of 1-allylterguride were separated by the use of mandelic acid as the selector [18]. The reduced errors in this reversed approach enabled a better estimate for the equilibrium constants between the enantiomers of mandelic acid and 1-allylterguride. Changing from D to L mandelic acid as the chiral selector enabled the migration order of the 1-allylterguride enantiomers to be reversed.

Another alkaloid which has been reported in the literature for the separation of enantiomers is *d*-(+)-tubocurarine (Figure 6.8) [19]. Tubocurarine is a neuromuscular blocking agent and is one of the alkaloids found in curare – which is famous for its use in the poison arrows traditionally employed in hunting in parts of South America.

The alkaloid *d*-(+)-tubocurarine was investigated for the separation of the enantiomers of a range of analytes containing carboxylic acid groups, including ketoprofen and the dansyl derivatives of some amino acids. The resolution values obtained from phosphate buffer systems (pH 5, 6, or 7) containing 15 mM of *d*-(+)-tubocurarine ranged from 0.6 to 2.05.

6.5 Crown Ethers

The application of crown ethers to the separation of enantiomers in CE began in the early 1990s with the use of 18-crown-6 tetracarboxylic acid [20, 21] as a buffer additive. The crown ether 18-crown-6 tetracarboxylic acid (Figure 6.9) has four carboxylic acid groups attached to asymmetric centres and these provide the possibility for enantioselective interactions.

The separation of the enantiomers of protonated primary amines by the crown ethers is thought to proceed via the formation of co-ordination complexes. It is believed that the six oxygen atoms in the crown ether form a planar arrangement and form three hydrogen bonds with the three protons attached to the protonated primary amine. When secondary and tertiary amines analytes are examined instead of primary amines no enantioseparation or complexation has been observed.

Twenty five amines including amino acids, small peptides, and biologically important amines such as norephedrine and noradrenaline were analysed in a 10 mM Tris/citrate buffer system at pH 2.2 containing 10 mM 18-crown-6 tetracarboxylic acid [20]. The crown ether displayed measurable enantioselectivity for twenty two of the amines with resolution values ranging between 0.54 (histidine) and 3.75 (phenyalanine). The enantiomers of ephedrine (a secondary amine) were not separated and the migration time indicated only very weak interaction with 18-crown-6 tetracarboxylic acid. The selectivity (as measured by the ratio of migration times) for the enantiomers of naphthylethylamine, tryptophan, glycylphenyalanine, and phenylalanine was measured as a function of temperature. For three of the analytes the selectivity decreased with temperature but for glycylphenyalanine it increased.

The performance of 18-crown-6 tetracarboxylic acid and α-cyclodextrin were compared in the separation of the enantiomers of phenyalanine, tyrosine, tryptophan, and dihydroxyphenylalanine [21].

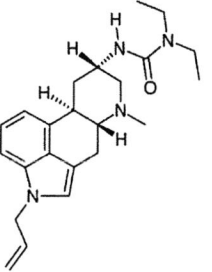

Figure 6.7. 1-allylterguride.

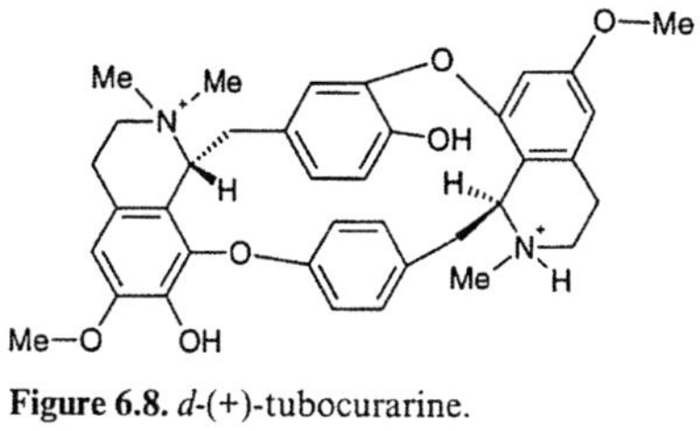

Figure 6.8. *d*-(+)-tubocurarine.

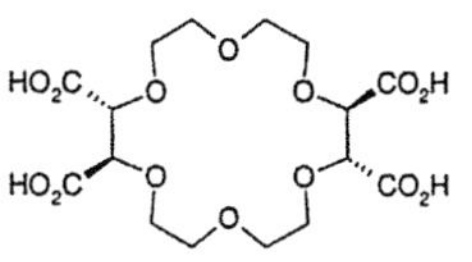

Figure 6.9. 18-crown-6 tetracarboxylic acid.

The best resolution was obtained by 18-crown-6 tetracarboxylic acid in two cases and by α-cyclodextrin in the other two cases.

The literature on the use of 18-crown-6 tetracarboxylic acid for the separation of enantiomers in CE has been reviewed and procedures given for synthesis, purification, and characterisation [22]. The use of 18-crown-6 tetracarboxylic acid for the resolution of amino acid enantiomers and some of their ester derivatives has been studied [23]. All of the amino acids except aspartic acid and glutamic acid could be resolved by 18-crown-6 tetracarboxylic acid. The lack of resolution of aspartic and glutamic acid enantiomers was thought to be due to electrostatic repulsion between the acid groups on the amino acids and those on the crown ether. The enantiomers of the acidic amino acids can be resolved using 18-crown-6 tetracarboxylic acid if they are converted into their esters. For example the enantiomers of the benzyl ester of aspartic acid were easily resolved using 10 mM of 18-crown-6 tetracarboxylic acid in a buffer a Tris/citrate buffer at pH 2.3. Interestingly the position of esterification made a large difference to both the migration times and the resolution: the enantiomers of the β-benzyl ester of aspartic acid (Figure 6.10) had a resolution value of 6.00 whereas

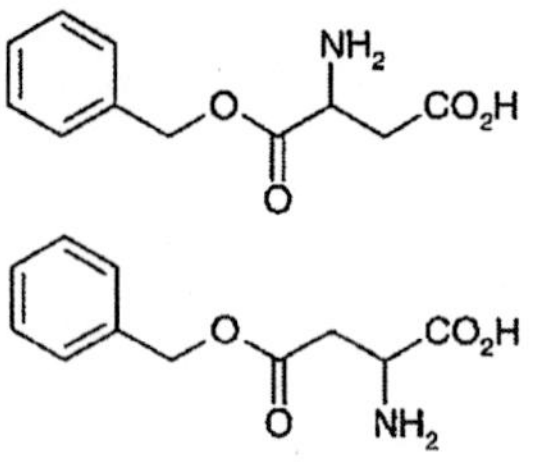

Figure 6.10. (*R, S*) aspartic acid α-benzyl ester and (*R, S*) aspartic acid β-benzyl ester.

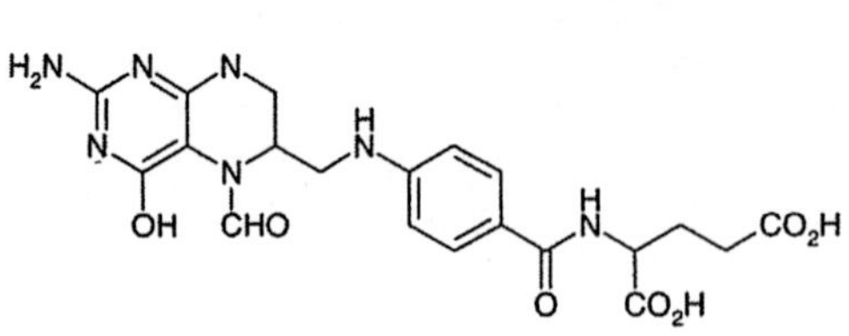

Figure 6.11. Leucovorin.

those of the α-benzyl ester had a resolution value of 1.97.

The formation of ester and other derivatives such as amides was found to have a major influence on the resolution of the enantiomers of some of the amino acids. In some cases the enantioresolution was reduced to zero and in other cases resolution values above ten were observed.

6.6 Natural Macromolecules

One of the most interesting groups of chiral selectors used in the separation of enantiomers by CE are natural macromolecules such as proteins, polysaccharides, and the macrocyclic antibiotics. Proteins in particular seem to be fitting selectors for the separation of enantiomers, given that a lot of the interest in the field has arisen because of the realisation that enantiomers can bind differently with proteins. If the two enantiomers can interact in different ways with living systems why not use some of the components of those living systems to effect enantiomer separation?

6.6.1 Proteins

The use of proteins as buffer additives in CE is an extension of their application in HPLC where they are commonly used for enantiomer separation both as stationary phases and to a much lesser extent as mobile phase additives. Whilst protein HPLC stationary phases can sometimes give very high selectivity they are often perceived as being expensive, giving low efficiency, and to lack robustness and reproducibility. Using proteins as buffer additives in CE is appealing because of the possibility to gain the selectivity advantages with fewer of the disadvantages. For example the use of conditions which might irreversibly denature the protein is less of a concern if the only impact is on the small amount of replaceable protein in a capillary rather than an expensive column. Section 6.6.1

contains a selection of the examples of the use of different proteins to separate enantiomers by CE.

Another important application of CE is as a way of studying the binding between drug development candidates and important human proteins. Because only small amounts of protein are required for CE it is possible to measure binding under a wide range of conditions e.g. different pHs, buffer salts and ionic strengths. Human serum albumin and α_1-acid glycoprotein are of particular interest as they are the principle plasma proteins associated with binding to drugs in plasma.

Serum albumins have been used as stationary phases in HPLC and were used relatively early as buffer additives in CE. For example Bovine Serum Albumin (BSA) was used to separate the (6*R*) and (6*S*) enantiomers of leucovorin [24] shown in Figure 6.11 (the 6 position is on the pteridine ring).

The leucovorin enantiomers were separated using a 1 mg/mL solution of BSA in phosphate buffer at pH 7. At pH 7 both leucovorin and BSA are negatively charged with leucovorin having the greater electrophoretic mobility. Interaction with BSA reduces the apparent electrophoretic mobility of the leucovorin enantiomers and the (6*R*) enantiomer was shown to have the greater affinity for BSA. Using a fused silica capillary lead to variability in migration times and poor capillary lifetimes – presumably due to protein adsorption. The problems of reproducibility and robustness were greatly reduced by the use of a capillary coated with polyethylene glycol. The resolution was found to increase with BSA concentration and both efficiency and resolution varied with the buffer pH.

Human Serum Albumin (HSA) has been used as a buffer additive to separate the enantiomers of kynurenine, tryptophan, 3-indole lactic acid, 2,3-dibenzoyl tartaric acid, and 2,4-dinitrophenyl glutamic acid [25]. The enantiomers were separated using a range of HSA concentra-

tions, and with pH 8 or 9 buffers (the enantioselectivity of albumin HPLC columns for carboxylic acids is highest near pH 9). With a freshly prepared 1 mg/mL solution of albumin in a buffer at pH 9.6 the enantioselectivity was found to vary over a three hour period. With analytes containing two carboxylic acid groups the enantioselectivity increased with time, whereas with analytes containing only one carboxylic acid group the enantioselectivity decreased. With the dicarboxylic acids constant enantioselectivity could be obtained by heating the albumin solution at 60 °C for 30 min before use, whereas for the monocarboxylic acids heating lead to a loss in enantioselectivity. The enantioselectivity of heat treated Human Albumin (HA) for 2,3-dibenzoyl tartaric acid (DBT) and 2,4-dinitrophenyl glutamic acid (DNPG) was measured at HA concentrations between 0 and 10 mg/mL. The mobility difference between the enantiomers of DBT and DNPG was strongly dependent upon the HA concentration. With DNPG the mobility difference increased in a non linear fashion over the measured concentration range and appeared to tend towards a limiting value. With DBT the mobility difference varied with concentration in a similar fashion to that seen with cyclodextrins (Chapter 3): the mobility difference increased with concentration up to a maximum value and then decreased slowly. Higher HA concentrations also gave sharper peaks but these were accompanied by greater baseline instability. The peak efficiencies measured in CE were about one order of magnitude higher than the values obtained by using HA as a stationary phase in HPLC.

One of the problems with the use of proteins as chiral selectors in CE is their high background UV absorbance. Proteins absorb strongly in the spectral range that contains the UV chromophores of typical analytes. An approach to solving the problem of high background absorbance is the partial separation zone technique [26, 27]. With this technique the protein containing buffer solution is used to fill the capillary up to a point before the detection region. The vials which are then used in the separation step contain only the buffer and no protein. Experimental conditions are chosen such that the proteins do not migrate through the detector and so do not contribute to the background. A typical set up involves the use of a coated capillary to eliminate electroosmotic flow, and pH conditions such that the analytes and proteins carry opposite charges and hence migrate in opposite directions. In this way the enantiomers migrate through the separation zone and then are detected in a zone which does not contain the protein.

The partial filling approach was used to separate the enantiomers of sixteen bases using BSA, conalbumin, ovomucoid, and α_1-acid glycoprotein [27] as the selectors. The buffer pH was chosen according to the isoelectric point of the protein to ensure a small overall negative charge on the protein. With a separation system contained 0.75 mM of BSA in a 50 mM phosphate buffer at pH 6, the measured resolution values for the bases ranged from 0.8 for trimebutine to 3.1 for the enantiomers of epinastine (Figure 6.12). Validation work showed good reproducibility for the migration times and peak areas, and good linearity for epinastine in the concentration range 10–100 µg/mL. The use of a range of BSA concentrations enabled the equilibrium constants to be estimated at 830 and 910 M^{-1} for (+) and (–) epinastine respectively.

With the partial separation zone approach the separation between the enantiomers can be altered by either changing the selector concentration or the length of the separation zone. For example the selectivity of human α_1-acid glycoprotein for the enantiomers of disopyramide was studied by varying the time used to fill the capillary with the protein solution [28]. The ratio of the migration times for the two enantiomers was directly proportional to the length of the separation zone.

The purity of the protein used is an important consideration as many proteins isolated from natural sources are in fact a mixture of components which may each display different enantioselectivities. An example is the resolution of the enantiomers of tolperisone by ovoglycoprotein (OG), a protein isolated from the crude ovomucoid (OM) extracted from chicken egg whites [29]. The crude OM was sepa-

rated into OG (about 10% w/w) and OM by cation-exchange chromatography and the enantioselectivity of the two fractions compared. The tolperisone enantiomers were partially resolved by crude OM, totally resolved by OG and unresolved by OM. These results indicate that the enantioselectivity shown by crude ovomucoid is due to the ovoglycoprotein content. The OG was also used for the separation of the enantiomers of other basic drugs such as verapamil and chlorpheniramine. With chlorpheniramine the enantioresolution was shown to increase between pH 4.5 and 5.5. The effect of organic solvent in the buffer was studied by the addition of different volumes of 2-propanol, ethanol and methanol. The organic solvents were shown to improve the peak shape (in particular that of the more strongly binding enantiomer) but to lead to a reduction in enantioselectivity.

Proteins are large molecules so in comparison to chiral selectors such as cyclodextrins there are many potential binding sites for small drug molecules. In addition it seems unlikely that all of the drug binding sites on proteins will be enantioselective. The issue of protein binding sites has been studied by the examining the enantioselectivity of avidin for a range of aromatic carboxylic acids including flurbiprofen and 4-fluoromandelic acid (Figure 6.13) [30, 31].

Avidin is a basic glycoprotein isolated from egg white which has an isoelectric point in the range 10 to 10.5. Avidin is well known because of its extremely high affinity for the low molecular weight compound biotin. The enantioseparation potential of avidin at concentrations between 0.05 and 0.1 mM in buffers from pH 4.5 and 6 was determined using fourteen acids and found to give resolution values between 1.2 and 3.0. The hypothesis that the four biotin binding sites on avidin were responsible for the enantioselectivity was investigated by adding different amounts of biotin to an avidin containing buffer [30]. As biotin has an extremely high affinity for avidin it could be expected to totally block the biotin binding sites to the acids. The separation between the enantiomers of 4-fluoromandelic acid decreased with increasing concentration of biotin and was lost entirely when the mole ratio of avidin to biotin was 1:4. Whilst all enantioselectivity had been lost by the addition of biotin the migration time of 4-fluoromandelic acid was still different from that seen in the absence of avi-

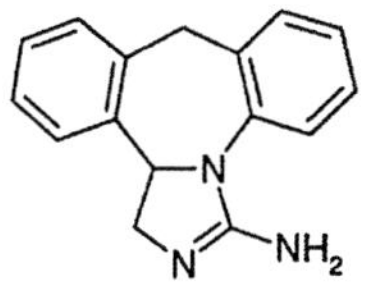

Figure 6.12. Epinastine.

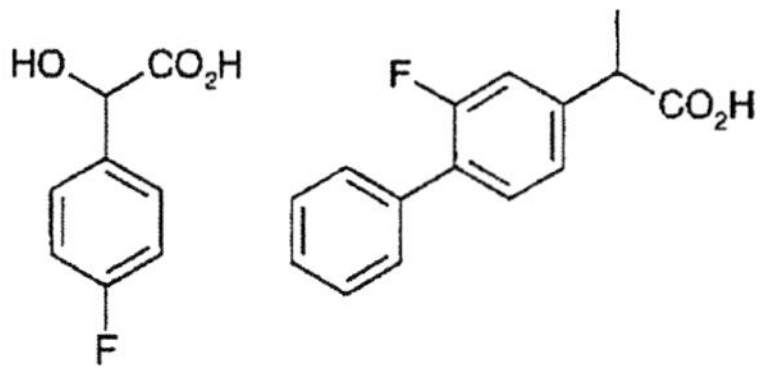

Figure 6.13. 4-fluoromandelic acid and flurbiprofen.

din. The results indicated that whilst the enantioselectivity of avidin is due to the biotin binding sites, other non enantioselective binding sites are also present. The strength of this non enantioselective binding appears to vary with the nature of the analyte.

6.6.2 Macrocyclic Antibiotics

The macrocyclic antibiotics were introduced into widespread use as chiral selectors in chromatography and CE in 1994 by Armstrong and co-workers [32–34]. The macrocyclic antibiotics employed cover several chemically distinct groups of species ranging in molecular weight from about 600 to about 2,200. The macrocyclic antibiotics have several stereogenic centres and the many different functional groups that they contain give the possibility for several different types of bonding interaction with analytes.

Vancomycin is one of the most widely used macrocyclic antibiotics selectors in CE and is well known because of the important role it plays as an antibiotic in human medicine. Vancomycin is an amphoteric glycopeptide containing three fused rings and has a relative molecular weight of 1449, its structure is shown in Figure 6.14 (R = H).

Vancomycin is produced by the soil bacterium *Streptomyces orientalis* and acts to prevent bacterial cell growth by binding to cell wall proteins. CE has been used as a method of investigating this inhibition process by measuring the binding between vancomycin and different peptide sequences [35, 36]. Vancomycin was found to bind much more strongly to peptides terminating in the sequence D-Ala-

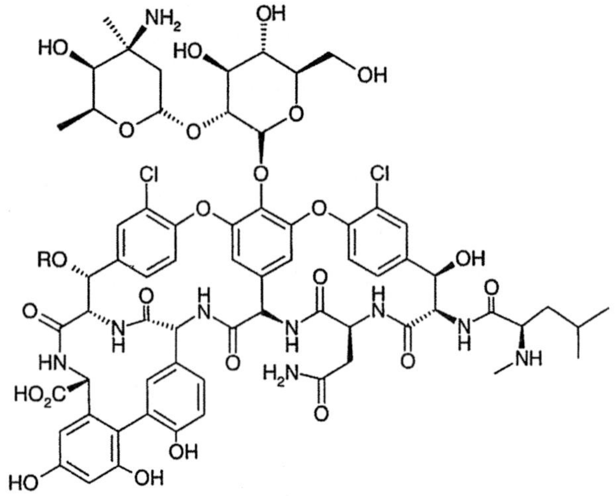

Figure 6.14. Vancomycin.

D-Ala than peptides where one of the alanines had been replaced by another amino acid, or one or both of the alanines had been inverted.

The potential of vancomycin as a chiral selector in CE was examined by studying the resolution between the enantiomers of over 100 carboxylic acids [32]. The analytes included amino acids which had been derivatised with a range of agents (e.g. dansyl, AQC, and PHTH derivatives) several non-steroidal anti-inflammatory drugs and other aromatic carboxylic acids. Very high enantioselectivities were observed and the high separation efficiencies obtained lead to high resolution. For the eight anti-inflammatory drugs examined the resolution obtained using 5 mM of vancomycin in a pH 7 buffer varied from 2.2 for indoprofen to 8.2 to flurbiprofen. Resolution values of 3 and above were obtained for seven of the analytes. High resolution values were also obtained for the derivatised amino acids with a value of nearly twenty being obtained for the dansyl derivative of methionine. The influence of buffer pH on enantioseparation was examined as pH can change both the charge on the analyte and that on vancomycin. Vancomycin contains both acidic and basic functional groups and the isoelectric point is estimated to be 7.2 in 100 mM phosphate buffer. The resolution between the enantiomers of the dansyl and AQC (6-aminoquinolyl-N-hydroxy-succinimidyl carbamate) derivatives of the amino acids was measured at both pH 4.9 and pH 7. With the AQC derivatives the resolution increased with pH whereas with the dansyl derivatives it decreased. Measurement of the resolution between

the enantiomers of naproxen, dansyl valine, and iopanoic acid at pH 5, 6, 7, and 8 showed a decrease in resolution with increasing pH. Optimising the pH is likely to be complicated as pH controls the charge on the analyte, the size and sign of the charge on vancomycin and the electroosmotic flow and all of these will have an impact on resolution. The importance of vancomycin concentration was determined by measuring resolution between the enantiomers of naproxen, dansyl valine, and iopanoic acid using vancomycin concentrations of 1, 2, and 5 mM. In all three cases the resolution increased – both because of an increase in the apparent electrophoretic mobility difference between the enantiomers and a reduction in the electroosmotic mobility (probably due to adsorption of vancomycin on the capillary walls). The low vancomycin concentrations used meant that direct UV absorbance detection was possible despite the absorbance of vancomycin.

In some cases the very high enantioselectivity of vancomycin arises because whilst one enantiomer has a high affinity for vancomycin the other enantiomer binds only very weakly. An example of this large difference in binding affinities is that of the AQC (6-aminoquinolyl-N-hydroxysuccinimidoyl carbamate) derivatives of the amino acids methionine, selenomethionine, and ethionine [37]. Whilst the D enantiomers bind strongly to vancomycin the electrophoretic mobilities of the L enantiomers are little changed in the concentration range 0.1–5 mM of vancomycin. These large affinity differences translate into apparent electrophoretic mobility differences between the enantio-

mers which are greater than 20% of the apparent electrophoretic mobility of the L enantiomer. These large mobility differences produced by vancomycin compare favourably with the values obtained with cyclodextrins. For example with propranolol and methyl-β-cyclodextrin the maximum mobility difference was about 2% of the electrophoretic mobility of the individual enantiomers (Chapter 3). Very high enantioselective binding has also been observed for vancomycin and several nonsteroidal anti-inflammatories, and dansylated amino acids [38]. Again whilst the electrophoretic mobility of one enantiomer changed markedly in the presence of a low concentrations of vancomycin, that of the other enantiomer hardly changed. The inherent enantioselectivities, (ratio of the equilibrium constants for the two enantiomers, K_1/K_2), were measured and some very high values obtained. The inherent enantioselectivities for dansyl valine and dansyl-α-n-butyric acid for example were 11 and 15.6 respectively.

The performance of vancomycin was compared with that of two derivatised β-cyclodextrins in the resolution of the enantiomers of some arylproprionic acids [39]. The resolution between the enantiomers of eight non steroidal anti-inflammatory drugs was compared using vancomycin; 2,3,6-tri-o-methyl-β-cyclodextrin; and heptamethylamino-β-cyclodextrin. In a pH 5 buffer vancomycin gave the highest enantiomeric resolution for six of the eight analytes, with each of the cyclodextrins giving the best resolution for one analyte each. The partial filling method was employed with vancomycin to improve the baseline stability (at pH 5 vancomycin is positively charged and so migrates in the opposite direction from the negatively charged analytes). The highest efficiency values and lowest detection limits were also obtained with vancomycin.

Vancomycin was also used to separate the enantiomers of ibuprofen and etodolac (Figure 6.15) and some of their metabolites [40]. The enantiomers of the drugs and their metabolites were detected by coupling electrospray mass spectrometry (MS) to the end of the CE capillary. The use of MS instead of UV absorbance for detection brings the advantages of helping to identify unknowns and helping to confirm that the peaks seen are due to enantiomers (same molecular weight and mass spectra). The use of selected ion monitoring enabled the spatially overlapping peaks due to ibuprofen enantiomers (m/z

= 205) to be resolved from those due to 2-hydroxyibuprofen (m/z = 221). The pH of 4.8 and the other analysis conditions were chosen such that the vancomycin moved away from the mass spectrometer and so did not contaminate the ion source.

Macrocyclic antibiotics with a similar structure to vancomycin have also been used as chiral selectors in CE. For example A82846B a glycopeptide antibiotic which has a similar structure and similar activity to vancomycin was used to resolve the enantiomers of three non-steroidal anti-inflammatory drugs and three dansylated amino acids [41]. A82846B differs structurally from vancomycin (Figure 6.14) in that it is epimeric at the sugar amino group and that the R group which is H in vancomycin is replaced by another epivancosamine group (Figure 6.16). The additional amino sugar means that the pI value of A82846B is higher than that of vancomycin.

The analytes were examined in 100 mM phosphate buffers at pH 5 and 6 using a partial filling approach and a coated capillary to ensure that the A82846B migrated away from the detector. The resolution between the enantiomers of the analytes increased on increasing the A82846B concentration from 1 to 2 mM. The resolution values at the higher concentration ranged from 2.8 to 13.1. For four of the analytes the enantioresolution was higher at low pH and for two it was lower.

The glycopeptide antibiotic ristocetin A which is structurally related to vancomycin has also used for the separation of enantiomers by CE [42]. Ristocetin A has a similar peptide backbone to vancomycin but has four macrocyclic rings and six sugar groups, as opposed to the three macrocyclic rings and two sugar groups of vancomycin. Ristocetin has a relative molecular weight of 2066, a similar pI value to vancomycin, and owes its antibacterial activity to binding to the same D-Ala-D-Ala peptide sequence. Ristocetin was used to resolve the enantiomers of over 120 analytes including non-steroidal anti-inflammatory drugs, amino acid derivatives, and carboxylic acids [42]. The analytes were resolved using 2 or 5 mM solutions of ristocetin using 100 mM phosphate buffers at pH 6 or 7. In general the enantioselectivities of ristocetin A and vancomycin are similar with some very high resolution values being obtained. The influence of the ristocetin concentration, buffer pH, and 2-propanol content of the buffer was examined by examining

the resolution between the enantiomers of 2-(3-chlorophenoxy) propionic acid, ketoprofen, 3-methoxymandelic acid, and 1-benzocyclobutenecarboxylic acid. In comparison to the standard conditions of 2 mM ristocetin, pH 6, and no propanol, the resolution decreased by increasing the pH and increased by increasing the ristocetin or 2-propanol concentrations. Ristocetin A also showed less tendency to adsorb to the walls of fused silica capillaries than vancomycin and is more stable in aqueous solution.

The antibiotic teicoplanin is also structurally related to vancomycin having a similar peptide backbone but with four macrocyclic rings and three sugar groups as opposed to three and two respectively [43]. With teicoplanin two of the sugar groups are acylated and differences in the aliphatic side chains on one of the sugars means that teicoplanin is normally a mixture of five main components [44]. In fact CE itself can be used to separate the main components in teicoplanin [45]. Teicoplanin is a surfactant with a critical micellar concentration in water of about 0.18 mM [43]. The enantioselectivities of vancomycin, ristocetin, and teicoplanin were compared by determining the resolution between the enantiomers of twenty eight acidic analytes [43]. The twenty eight acids were analysed using 2 mM solutions of the antibiotics in 100 mM phosphate buffers at pH 6. Ristocetin gave the best resolution for seventeen of the analytes, with the scores for vancomycin and teicoplanin being nine and two respectively. Solutions of ristocetin also had the greatest stability whilst those of vancomycin had the least. The use of vancomycin almost always lead to the longest analysis times.

Teicoplanin was used to resolve the enantiomers and diastereoisomers of twenty one derivatised and underivatised di- and tripeptides [45]. High resolution values were obtained using 1.2 mM teicoplanin in a Tris/phosphate buffer at pH 6.25 containing 40% acetonitrile, with the maximum being 37.8 for Gly-Ala, and values over ten were recorded for nine of the analytes. Teicoplanin, as with vancomycin, was shown to bind very enantioselectively to some of the peptides by examining the change in separation produced by concentrations of teicoplanin in the range 0 mM to 1.6 mM. For example whilst the migration times of the L enantiomers of Gly-Asp and Ala-Gly were essentially unchanged by increasing the concentration of teicoplanin, those of the D enantiomers

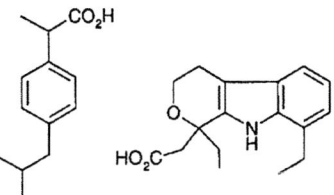

Figure 6.15. Ibuprofen and etodolac.

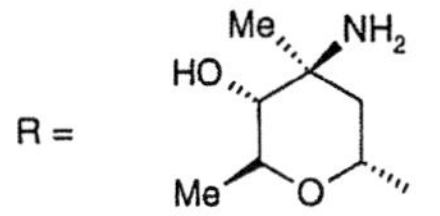

R =

Figure 6.16. epi-vancosamine.

changed markedly. The nature of the derivatisation reagent coupled to the peptides also had a significant impact upon the resolution at the single teicoplanin concentration examined.

The influence of the teicoplanin concentration, buffer pH, and buffer concentration was determined by measuring the resolution between the enantiomers of mandelic acid and those of three mandelic acid derivatives under different conditions [44]. Increasing the teicoplanin from 0.1 mM to 0.3 mM increased the electrophoretic mobility difference and enantioresolution for all of the analytes. Increasing the buffer concentration from 25 mM to 100 mM at pH 4 increased enantioresolution. The relationship between buffer pH and enantioresolution was more complex with change from pH 4 to 5 leading to an increase in resolution, but further pH increases leading to a continual decrease. Blocking the D-Ala-D-Ala binding site in teicoplanin results in a loss of enantioresolution demonstrating the role of this binding site in discrimination between enantiomers. The resolution between the enantiomers of N-benzoyl-alanine, and 4-methoxymandelic acid was measured in buffer solutions containing different concentrations of either N-acetyl-D-Ala-D-Ala or N-acetyl-L-Ala-L-Ala. Increasing concentrations of N-acetyl-D-Ala-D-Ala lead to a loss of resolution, whereas N-acetyl-L-Ala-L-Ala did not result in any change.

The macrocyclic antibiotic rifamycin B has been used for the separation of the enantiomers of basic analytes in CE [33]. Rifamycin B gives better results with basic rather than acidic analytes and so can be regarded as complementary to the vancomycin family which give best results with acidic analytes. The rifamycins are isolated from the bacterium *Norcardia medi-*

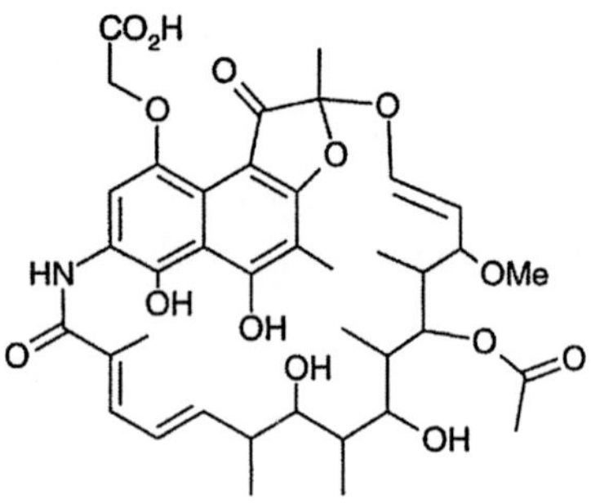

Figure 6.17. Rifamycin.

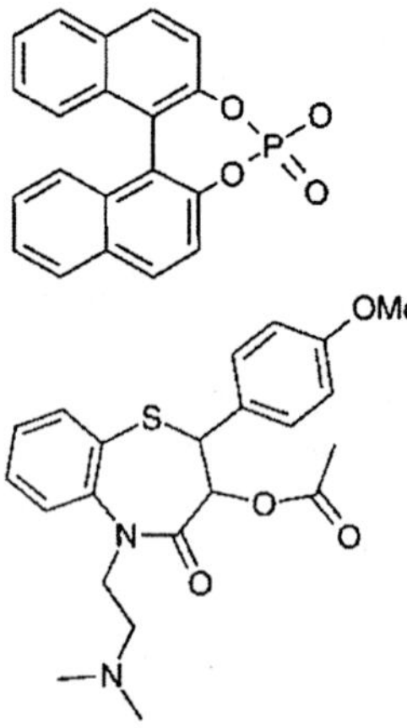

Figure 6.18. 1,1'-binaphthyl-2,2'-diyl hydrogen phophate and cis-diltiazem.

terranei and act by the inhibition of DNA polymerase [33]. The rifamycins belong to a different family of antibiotics to vancomycin and the structure of rifamycin B is shown in Figure 6.17.

A 25 mM solution of rifamycin B was used in a buffer composed of 40% 2-propanol/60% 100 mM phosphate at pH 7 to resolve the enantiomers of eighteen pharmacologically active amino alcohols including β-blockers, bronchodilators, and andregenics. The resolution values ranged from 0.4 for oxprenolol to 3.1 for terbutaline. The analytes were monitored by inverse detection at 254 nm because of the background absorbance caused by the 25 mM solutions of rifamycin. The importance of the buffer parameters was determined by measuring the resolution between the enantiomers of terbutaline, norphenylephrine, and metoprolol using different conditions. The resolution increased as the rifamycin concentration changed from 15 to 25 mM, with the optimum pH being about 7. The type and concentration of organic solvent employed changed the resolution with maximum values being obtained using 40% *v/v* of 2-propanol.

6.6.3 Polysaccharides

Polysaccharides such as celluloses have enjoyed some success as chiral stationary phases in HPLC and both polysaccharides and oligosaccharides have received attention as chiral buffer additives in CE. For example maltodextrins were investigated as chiral selectors for some nonsteroidal anti-inflammatory drugs and warfarin and some of its derivatives [46]. The maltodextrins are linear polymers of D-glucose which are joined through their 1 and 4 positions by α linkages. Maltodextrins of different chain lengths and from different sources were evaluated using high concentrations in the range 2.5 to 10% *w/v*. For Ibuprofen and flurbiprofen baseline resolution was achieved with the degree of resolution increasing with the maltodextrin concentration and chain length. A detailed understanding of the relationship between chain length and resolution was not achievable because of the difficulty in fractionating maltodextrins above a chain length of seven units. Baseline resolution was also achieved with warfarin and its derivatives.

A range of natural and derivatised polysaccharides was investigated for the separation of the enantiomers of 1,1'-binaphthyl-2,2'-diyl hydrogen phosphate (BDHP) and some other analytes [47]. Baseline resolution of BDHP, trihexyphenidyl, and mianserine is obtained by using high concentrations of the polysaccharides (up to 20 mg/mL). The enantioselectivity of the polysaccharides was found to vary according to their structure. Sugars with α-(1-4) and α-(1-6) linkages (amylose and pullulan) bound preferentially to (R)-(–) BDHP, and sugars with β-(1-4) and β-(1-6) linkages (laminaran, methyl cellulose, and hydroxypropyl cellulose) bound preferentially to (S)-(+) BDHP. The resolution between the enantiomers of BDHP and *cis*-diltiazem (Figure 6.18) varied with the concentration of amylose-6250 (average chain length 6250 units) in the buffer. Graphs of resolution vs. concentration had a similar form to those observed with β-blockers and cyclodextrins (Chapter 3) with the optimum amylose concentration being 15 mg/mL.

6.7 Non-Aqueous CE

Most separations of enantiomers in CE are carried out in buffer systems in which water is the only or the main solvent, and

in this respect the trend follows that observed with CE generally. Non-aqueous solvents, however, can also be used for enantioseparation in CE and this is an area of recent and growing interest. The potential benefits of non-aqueous solvent systems can be grouped into two areas: the application to analytes which are not sufficiently soluble in aqueous based buffer systems, and the use of different solvents to offer the possibility of different routes to obtaining enantioselectivity.

A general aim in the choice of conditions for the separation of enantiomers is that of maximising the enantioselective interactions between the analyte and the selector, whilst minimising the non-selective ones. Analytes can interact with the selector by a variety of means e.g. solvophobic, ion-pair, dipole-dipole, donor-acceptor etc., and the size and relative importance of these interactions will depend upon the properties of the bulk solvent [48]. For example electrostatic interactions are likely to be more significant in non-polar solvents than in water, and hydrophobic interactions less significant. The use of solvents such as formamide, *N*-methylformamide (NMF), *N,N*-dimethylformamide (DMF), methanol, and acetonitrile therefore offers the possibility of not only improving analyte solubility but also of changing enantioselectivity.

Cyclodextrins are the most commonly used chiral selectors in capillary electrophoresis and the separation mechanism is usually thought to proceed via the incorporation of a hydrophobic part of the analyte into the hydrophobic cyclodextrin cavity. If inclusion is primarily driven by hydrophobicity it could be expected that there will be less tendency to include in non-aqueous separation systems. This reduced tendency for inclusion has been measured in solvent systems containing the same electrolyte but with solvents of different polarity. For example the size of the binding constants between β-cyclodextrin and mianserin, trimipramine, and thioridazine (Figure 6.19) have been measured in water, 6 M aqueous urea, formamide, NMF, and DMF [49].

The binding constants in water are larger than the other solvents with the following values being obtained for the more strongly bound thioridazine enantiomer: water 3.54×10^4; 6 M aqueous urea 3.82×10^3; formamide 6.98; NMF 0.59; and DMF 5.9×10^{-2}. As a result of the large differences in binding constants the optimum β-cyclodextrin concentration is very different in the different solvents. In aqueous solution the thioridazine enantiomers were resolved with concentrations of β-cyclodextrin in the range 0.04–0.5 mM; in 6 M urea with concentrations in the range 0.25–2 mM; and in formamide the β-cyclodextrin concentration had to be above 20 mM. The thioridazine enantiomers were not resolved in either NMF or DMF. The resolution between the enantiomers of eleven basic analytes was determined in formamide containing 150 mM of citric acid, 100 mM of Tris, and 100 mM of cyclodextrin. As with conventional CE the different cyclodextrins gave different performance: none of the analytes could be resolved with α-cyclodextrin, and the β-cyclodextrins (the parent and the methyl and hydroxypropyl derivatives) gave better enantioselectivity in general than γ-cyclodextrin.

Because of the relative weakness of the hydrophobic interactions, neutral cyclodextrins are not very attractive as chiral selectors in non-aqueous solvents. Charged cyclodextrins with a charge of opposite sign to that on the analyte are a more interesting proposition in non-aqueous systems because of the possibility of electrostatic interactions. For example the enantiomers of some dansylated amino acids and acidic anti-inflammatory drugs can be resolved using 20 mM of quaternary ammonium β-cyclodextrin, and an ammonium acetate/acetic acid electrolyte in formamide [50]. The resolution values for the anti-inflammatory drugs varied between 3.22 (fenoprofen) and 0.49 (indoprofen), and for the dansylated amino acids between 9.60 (valine) and 2.24 (phenylalanine).

In an analogous fashion the enantiomers of aromatic basic analytes can be resolved using heptakis(2,3-diacetyl-6-sulphato)-β-cyclodextrin (HDAS-β-CD) in methanol [51]. The enantiomers of thirteen basic analytes and one quaternary ammonium compound were resolved using HDAS-β-CD concentrations of either 10, 20, or 40 mM and an electrolyte prepared from 50 mM trichloroacetic acid and 25 mM triethylamine. The resolution increased with the concentration of HDAS-β-CD with the values for terbutaline and propranolol at 40 mM being 5.3 and 4.3 respectively. Acidic and neutral analytes only had very weak interactions with HDAS-β-CD.

In the previous discussion it was noted that hydrophobic interactions are likely to be less significant in non-polar solvents

than in water, and that electrostatic interactions are likely to be more important. For example strong electrostatic attraction between ions of opposite charges could be expected to lead to the formation of ion-pairs in non-polar solvents. Such electrostatic interactions are exploited in HPLC where chiral ion-pair reagents are used as mobile phase additives. Some of these HPLC chiral ion-pair reagents have also been successfully employed in non-aqueous CE for the separation of enantiomers. For example sodium or potassium camphorsulphonate (Figure 6.20) can be used to separate enantiomers of fifteen pharmaceutical analytes including some β-blockers and other β-amino alcohols [52].

The camphorsulphonates are dissolved at concentrations between 5 mM and 60 mM in acetonitrile containing 1 M acetic acid. Of the fifteen analytes examined only those which had a β-amino alcohol configuration could be separated. With salbutamol the maximum enantioresolution was obtained at a concentration of 30 mM (+)-(S)-camphorsulphonate. A concentration of 30 mM of (+)-(S)-camphorsulphonate in water instead of acetonitrile did not give any separation. With the acetonitrile and 1 M acetic acid system the resolution could be improved by the addition of between 0.1 and 0.2 mM Tween 20 to suppress the electroosmotic flow, and this approach gave baseline resolution for the enantiomers of atenolol and metoprolol. The migration order of the metoprolol enantiomers is reversed upon changing the ion-pair reagent from (+)-(S)-camphorsulphonate to (–)-(R)-camphorsulphonate. The addition of acetate to the separation system leads to a loss in enantioresolution – presumably due to the formation of competing ion-pairs.

Non-aqueous CE can also be used for the separation of enantiomers by the ligand exchange approach (compare Section 6.2). The enantiomers of eight amino acids were separated using the copper (II) complexes of L-proline and L-isoleucine dissolved in a solution of 25 mM ammonium acetate and 1 M acetic acid in methanol [53]. The separation depended upon the apparent pH of the electrophoresis medium (and so the charge on the analyte and the exchange ligand) and was optimised using a ratio of L-proline to copper (II) of about 3:1. The migration order of tryptophan enantiomers can be reversed by changing the exchange ligand from L-proline to D-proline. In contrast

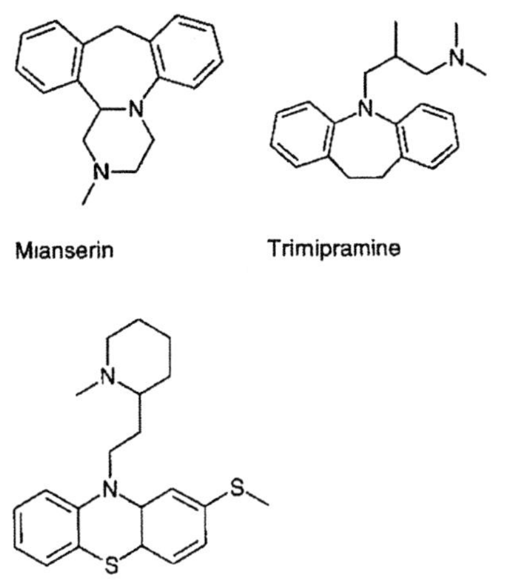

Mianserin   Trimipramine

Thioridazine

Figure 6.19. Mianserin, trimipramine, and thioridazine.

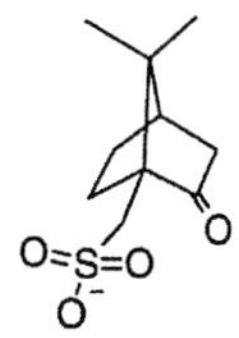

Figure 6.20. Camphorsulphonate.

to ligand exchange carried out under aqueous conditions, better separations could be obtained and at lower concentrations of the copper (II) complex.

6.8 Derivatisation

In many ways the approaches to the separation of enantiomers by CE have followed the examples set in chromatography and this is also the case with enantiomer derivatisation. Derivatisation reagents are used in CE to either improve the separation or detection steps. Derivatisation procedures have been most commonly applied in CE to the analysis of amino acid enantiomers where both of these steps can be problematic. Many amino acids lack a strong chromaphore and so are difficult to detect at low levels by UV absorbance detection. The formation of derivatives which have a strong UV chromaphore or are fluorescent can thus greatly assist the detection of low levels of amino acids and the use of dansyl and other derivatives are widely reported. The formation of derivatives can also improve the interaction of the enantiomers with the chiral selector e.g. hydrophobic interactions between aromatic derivatisation reagent and cyclodextrins. A recent

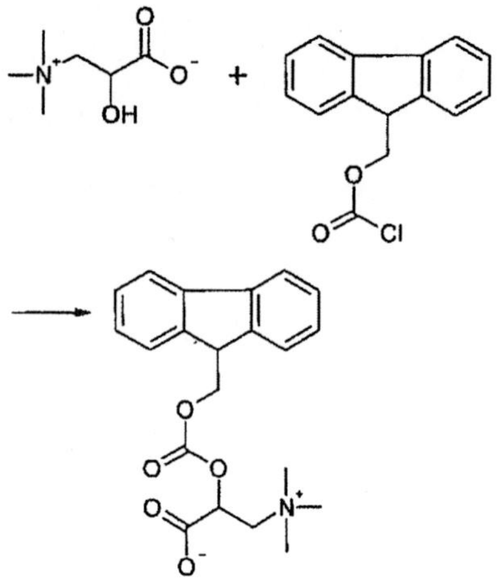

Figure 6.21. The derivatisation of carnitine with FMOC.

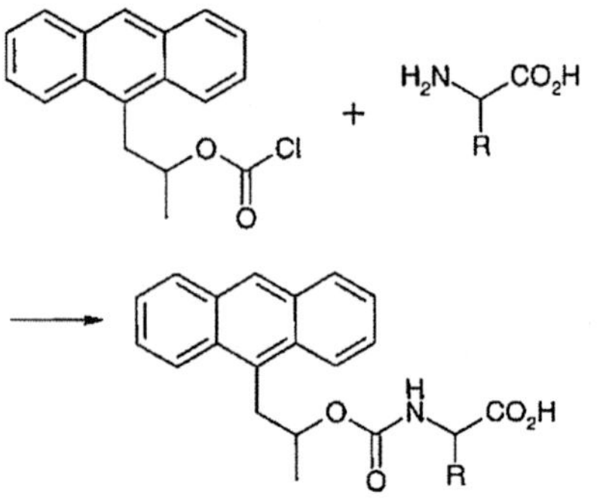

Figure 6.22. The reaction between APOC and amino acids.

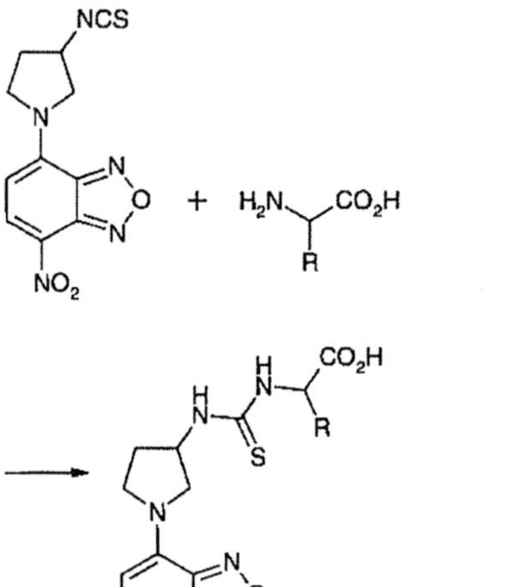

Figure 6.23. The reaction between NBD-PyNCS and amino acids.

example is the reaction of the enantiomers of carnitine with 9-fluorenylmethyl chloroformate (FMOC) (Figure 6.21) followed by separation of the enantiomers by 2,6-dimethyl-β-cyclodextrin [54]. In the absence of the derivatisation step the carni-

tine enantiomers could not be resolved. The derivatisation step enabled D-carnitine to be quantified in the presence of a hundredfold excess of L-carnitine.

The use of a derivatisation step does however complicate the analytical method and introduce additional issues which must be addressed in the validation of the method. For example some of the important issues are: the extent to which derivatisation speed and completeness depends upon the structure and concentration of the analytes, interference from the derivatisation reagent or its impurities, and the influence of the analyte matrix on derivatisation. A particular issue with fluorescent derivatives is the extent to which fluorescence depends upon environmental factors which can be different for the different analytes e.g. incorporation into cyclodextrins.

Another approach to the derivatisation step is the use of asymmetric derivatisation reagents to form diastereoisomers from the analyte enantiomers. The diastereoisomers can then be separated by standard CE methods without the need for the addition of a chiral selector to the buffer. With this approach the derivatisation step not only improves detection but also directly differentiates between the enantiomers. One of the disadvantages of the approach is that for quantitative analysis it is necessary to demonstrate that both enantiomers will react to the same extent with the derivatising reagent. In addition the optical purity of the derivatisation reagent must be significantly higher than that of the analyte or significant interference will be caused.

An example of enantiomer separation via diastereoisomer formation is the use of either (+) or (–)-1-(9-anthryl)-2-propyl chloroformate (APOC) to derivatise the enantiomers of amino acids and peptides (Figure 6.22) [55].

The APOC reagent was found to react at the same rate with the enantiomers of alanine and proline and to give a response which is linear in the concentration range 10 nM to 1 μM. The enantiomeric purity of the (+) and (–) APOC reagents was determined by reacting each enantiomer with glycine and monitoring the reaction products. The diastereoisomers of the amino acids were then separated using a borate buffer at pH 10 containing 10 mM of SDS. A racemic mixture of twenty amino acids was analysed by this approach and selectivity was seen for seventeen of them with some co-elution of the diastereoisomers of

the faster migrating amino acids. Increasing the SDS content of the buffer lead to improved separation between the faster migrating amino acids but at the expense of the slower ones. The diastereoisomers from the amino acids can be detected using either UV absorbance (256 nm) or Laser-Induced Fluorescence (LIF) detection (excitation 351 nm, detection 412 nm). The limit of detection by fluorescence is nearly four orders of magnitude lower than that achieved by UV absorbance. In each case the diastereoisomers formed from the L amino acids and (+) APOC had faster migration times than those formed from the D amino acids. This approach can also be used to determine the enantiomeric purity of small glycine containing peptides such as Ala-Gly and Ala-Gly-Gly. Levels of minor enantiomers below the 0.1% level could be estimated following correction for the enantiomeric purity of the derivatisation reagent.

Other asymmetric fluorescent derivatisation reagents have also been used for the analysis of amino acid enantiomers, for example the R and S enantiomers of 4-(3-isothiocyanatopyrrolidin-1-yl)-7-nitro-2,1,3-benzoxadiazole [R-(–) or S-(+)-NBD-PyNCS], (Figure 6.23) [56].

The diastereoisomers formed from the amino acid enantiomers are separated in a 25 mM acetate buffer at pH 4 containing 10 mM of the surfactant Triton X-100. The diastereoisomers formed from the amino acid enantiomers are monitored using LIF with an Argon ion laser (488 nm excitation, 520 detection). The diastereoisomers formed from the L amino acid enantiomers and S-(+)-NBD-PyNCS migrated faster than those formed from the D enantiomers. The migration order is reversed when R-(–)-NBD-PyNCS is used as the derivatisation reagent.

6.9 Capillary Electrochromatography

Another approach to the separation of enantiomers using CE equipment is via the technique of capillary electrochromatography (CEC) using enantioselective stationary phases. CEC differs from conventional HPLC in that the movement of mobile phase is driven by electroosmosis rather than by the application of pressure. With CEC the size of the electroosmotic flow is dominated by the zeta potential between the mobile phase and the stationary phase surface, rather than the mobile phase and the capillary wall, because of the much higher surface area. With enantioselective CEC the stationary phase performs the functions of providing both the driving force for the mobile phase movement and the enantioselective separation. Because of the dual stationary phase functions there are therefore more constraints on separation conditions in CEC than there are in HPLC. CEC was first demonstrated by Jorgenson and Lukacs [57] and has been the subject of a great deal of research since then. The interest in CEC arises because of the potential of the technique to produce higher efficiencies than are possible in HPLC. In HPLC the efficiency (sharpness of the analyte peaks) depends upon a number of factors related to the separation system and the analyte. One conventional measure of efficiency is the height (or length) of column equivalent to a theoretical plate, H. Highly efficient columns have small H values and vice versa. H is related to the linear velocity of the mobile phase and a number of different equations have been developed to model the relationship e.g. equation (6.1) [58].

$$H = A \cdot d_p + \frac{B \cdot D_m}{u} + \frac{C \cdot u \cdot d_p^2}{D_m} \quad (6.1)$$

Where u is the linear velocity, d_p is the diameter of the stationary phase packing, D_m is the diffusion coefficient of the analyte, and A, B, and C are constants.

The different terms in equation (6.1) arise because of the different physical contributions to band broadening in a chromatographic system. The first term arises because the flow through a packed bed is tortuous and not all of the paths are of the same length. The second term arise due to diffusion of the analyte zone along the length of the column. And the third term arises because of the resistance to mass transfer of the analyte between the stationary and mobile phases. Equation (6.1) indicates that the efficiency is highest with columns packed with small particles and that the influences of the linear velocity and analyte diffusion constants are complex.

Figure 6.24 shows experimentally determined values of H as a function of linear velocity for the β-blocker propranolol on a column packed with 5 micron C18 particles.

The plot in Figure 6.24 shows the composite effect of the three terms in equation (6.1). At very low flow rates H is large because of analyte diffusion along the length of the column. At very high flow rates H is

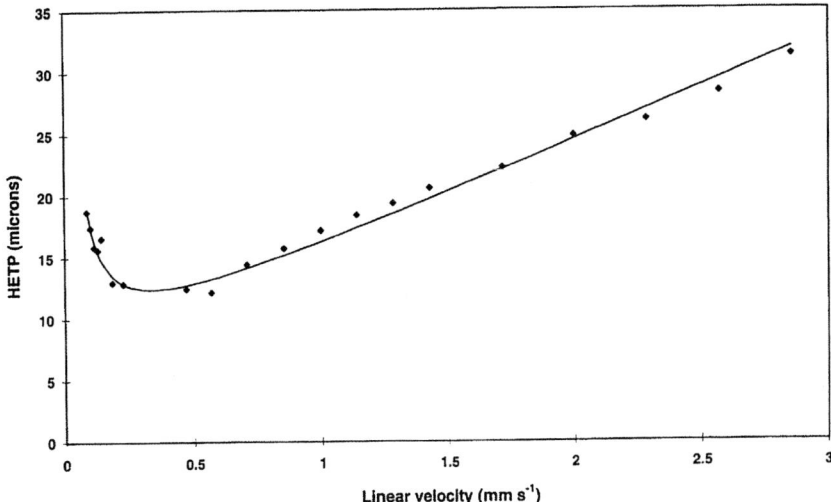

Figure 6.24. The experimental relationship between H and u.

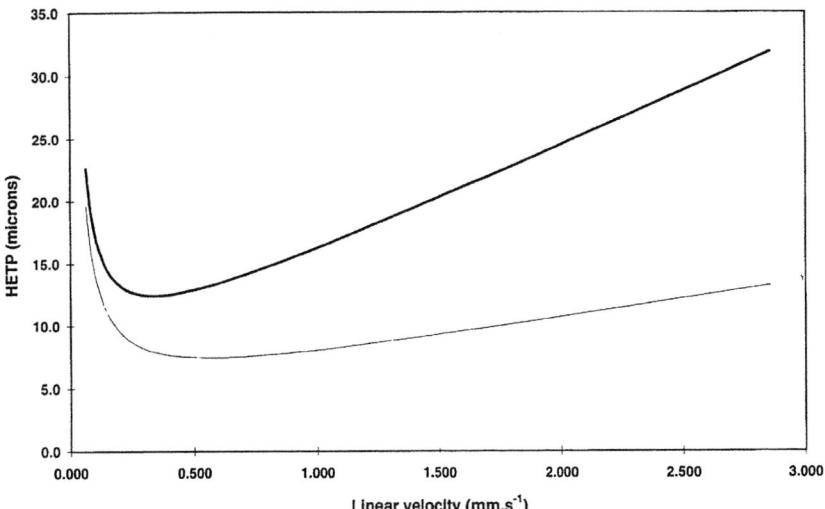

Figure 6.25. The influence of particle size on H for 5 micron particles (thick line), and 3 micron particles (thin line).

large because of the linear relationship between velocity and the mass transfer term. At intermediate flow rates there is a minimum in the curve, although in HPLC this minimum occurs at a linear velocity below the normal operational range. From equation (6.1) it follows that higher efficiency (lower H values) can be obtained by the use of smaller stationary phase particles. Figure 6.25 shows the curves generated using particle sizes of 3 and 5 microns.

The limit to using smaller particles in HPLC is the pressure drop across the column which increases with the inverse of the square of the particle size, equation (6.2), [58]. For example reducing the size of the stationary phase particles from 5 to 3 microns increases the pressure by a factor of 2.78.

$$\Delta p \propto \frac{\eta \cdot u \cdot L}{d_p^2} \quad (6.2)$$

where Δp is the pressure drop, L the length of the column, and η the mobile phase viscosity.

The interest in CEC as alternative to HPLC arises because of the desire to use smaller stationary phase particles without the penalty of very high operating pressures. In CEC the driving force for the movement of the mobile phase is the zeta potential between the stationary phase particles and the surrounding mobile phase. Experimental results for a range of stationary phase particle sizes indicate that the electroosmotic flow generated is independent of the particle size [59, 60]. This result means that smaller particles may be

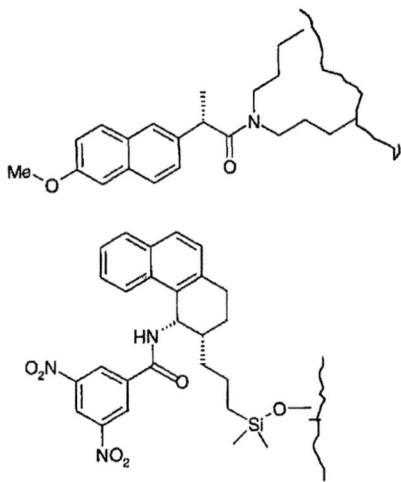

Figure 6.26. The (S)-Naproxen and (3R, 4S)-Whelk-O stationary phases.

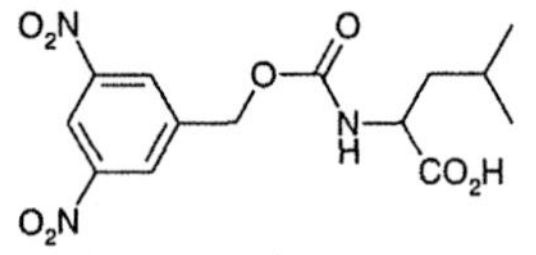

Figure 6.27. N-(3,5-dinitrobenzyloxycarbonyl) leucine.

employed in CEC than are possible in HPLC because of pressure restrictions. The promise of CEC is the possibility to use long columns packed with small stationary phase particles and so generate much higher operating efficiencies than are possible in conventional HPLC. Other results indicate that electrically driven flow has other benefits. Work on the same capillary columns using either pressure or electrically driven flow shows that significantly higher efficiencies are generated with electrically driven flow [59, 61]. The improvement in efficiency is due to reduction in both the A and C terms in equation (6.1) by factors of between two and four [61]. The improvements arise in part because of the flatter flow profile seen in electrically as opposed to pressure driven flow.

Whilst CEC has many potential advantages over HPLC the technique is far from routine. The lack of routine usage is in part due to a number of significant technical issues. CEC is normally performed on CE instruments and the analytes have to be injected electrokinetically because of the low pressures available with commercial equipment. It can be difficult to pack capillaries with very small particles and there are problems with robustness and reproducibility. A more fundamental issue is that the stationary phase particles

have two tasks to perform: the generation of the electroosmotic flow and the separation chemistry. These two tasks are not always compatible with one problem being that active silicas which give the highest zeta potentials also give poor results for basic analytes. The size of the electroosmotic flow is also an issue, with the linear velocities generated by even short CEC columns being significantly below those normally achieved in HPLC. The relatively low linear velocities lead to long analysis times and do not favour the practical application of very long columns. Because of these issues the volume of literature on enantiomer separation by CEC is smaller than might have been predicted ten years ago.

Most of the literature work on the separation of enantiomers by CEC involves the use of stationary phases and packings which have been successfully used in HPLC. An α_1-acid glycoprotein (AGP) stationary phase bonded to silica gel has been employed for the separation of the enantiomers of some β-blockers, barbituates and other compounds [62]. Nineteen analytes were examined in a mobile phase composed of phosphate buffer at pH 6.8 and low concentrations of either propanol, ethanol, methanol, or acetonitrile. The nineteen analytes had all been resolved into their enantiomers by HPLC but only ten were by CEC using the same phase. Of the remaining analytes most were acids which migrated electrophoretically in the opposite direction to the electroosmotic flow and were not detected – either because they were not eluted or could not be loaded onto the column. The efficiencies were in general higher than those seen by HPLC but the peaks still tailed significantly – presumably because of the properties of AGP itself.

A stationary phase composed of the antibiotic teicoplanin bonded to 5 μm silica has been used for the separation of the enantiomers of tryptophan and dinitrobenzoyl leucine [63]. The enantiomers of both analytes were baseline resolved in under six minutes using a mobile phase composed of 70% acetonitrile and 30% 2 mM phosphate buffer at pH 7. The enantioselectivity was affected by the organic solvent used, with ethanol and methanol giving higher selectivity than acetonitrile. The robustness of the column preparation procedure was determined by measuring the enantioresolution of tryptophan on seven columns. The average resolution value obtained from the seven

columns was 1.74 with an RSD of 11%. The comparable data for the repeatability of a single column are 1.70 and 0.5% respectively. The average efficiency value obtained from the seven columns was 29,000 plates m^{-1}.

Brush type chiral stationary phases have also been employed in CEC. For example 3 μm silica particles bonded with the (S)-Naproxen based phase and a (3R, 4S)-Whelk-O stationary phase (Figure 6.26) have been employed for the enantioresolution of some neutral analytes [64].

The neutral compounds were analysed using a mobile phase composed of 22% buffer at pH 6 and 78% acetonitrile. Under these mobile phase conditions good enantioresolution and high efficiencies were obtained with low retention (capacity factors of between 0.26 and 1.28). With the Naproxen based stationary phase the resolution values for five analytes varied between 8 and 17. With the Whelk stationary phase the values for five analytes varied between 2.6 and 31. Good peak efficiencies were obtained on both of the stationary phases with values of up to 200,000 plates m^{-1} being recorded. The high efficiencies are ascribed to the good mass-transfer characteristics of brush-type chiral stationary phases.

Anion-exchange chiral stationary phases have been employed for the separation of the enantiomers of some N-derivatised alpha-amino acids such as the 3, 5-dinitrobenzyloxy carbonyl derivative (DNZ) of leucine [65]. The enantiomers of the derivative (Figure 6.27) were separated using a weak anion exchange stationary phase based upon quinine which had been bonded to 5 μm silica particles.

The performance of the quinine based stationary phase was compared in both 25 cm × 100 μm CEC columns and 15 cm × 4.6 mm HPLC columns using the same mobile phase and linear velocity. The enantioselectivity (as measured by the ratio of the capacity factors) was the same for both the CEC and HPLC columns but the separation efficiency differed. With the CEC column the efficiency value for the least retained (R) enantiomer (120,000 plates m^{-1}) was twice as high as the values recorded using HPLC. For the (S) enantiomer however the CEC column efficiency was only two thirds of the HPLC values. In the CEC approach there are some additional complications to the analysis conditions which are absent in HPLC. Changing the buffer pH and ionic strength to alter retention also results in changes to the electroosmotic flow and so linear velocity in the CEC mode.

Whilst much of the literature devoted to the separation of enantiomers by CEC employs chiral HPLC stationary phases, a number of other approaches have been adopted. For example the cyclodextrins which have been widely used in CE may be immobilised to form a stationary phase by incorporation into polyacrylamide gels [66, 67]. The approach has been used to separate enantiomers of dansylated amino acids and shows high separation efficiency and good robustness and reproducibility. Another CEC enantiomer separation approach employs stationary phases produced by molecular imprinting [68]. In this example (R)-propranolol was used as a template to produce a stationary phase which was subsequently used for the separation of the enantiomers of propranolol and other β-blockers. The best enantioresolution was obtained for propranolol although values between 0.7 and 1.7 were obtained for the other β-blockers. As expected the (R) enantiomers were the most strongly retained although the peak shape was poor with extensive tailing.

Enantiomer separations can also be performed using open tubular CEC columns. For example capillaries coated with β- or γ-cyclodextrin bonded to dimethylpolysiloxane have been employed for the separation of the enantiomers of e.g. ibuprofen, flurbiprofen, 1-phenylethanol, and 1,1′-binaphthyl-2,2′-diylhydrogenphosphate [69, 70]. The 50 μm fused silica capillaries were coated with a sub micron layer of the stationary phase and the analytes were separated using phosphate buffers at pH 7. With 1-phenylethanol as the analyte, increasing the film thickness from 0.2 to 0.8 μm lead to an increase in the capacity factor from 0.21 to 1.10 but gave a drop in efficiency from 30500 to 9500. This drop in plate count is presumably due to poorer mass transfer with the thicker films. Increasing the separation voltage did not change the selectivity or capacity factor but lead to a reduction in the plate height with the minimum observed being 3.9 μm.

6.10 Conclusion

Chapter 6 has covered some of the selectors other than cyclodextrins which are used as buffer additives to separate enantiomers in free solution CE, and some of the alternative approaches to enantioseparation in electrically driven separations.

Whilst these other selectors and approaches are not as generally or as easily applicable as cyclodextrin buffer additives they offer real benefits for some analytes, and should not be dismissed.

References

[1] Gassman, E.; Kuo, J.E.; Zare, R.N. Electrokinetic Separation of Chiral Compounds, *Science* **1985**, *230*, 813–814.

[2] Gozel, P.; Gassman, E.; Michelsen, H.; Zare, R.N. Electrokinetic Resolution of Amino Acid Enantiomers with Copper (II)-Aspartame Support Electrolyte, *Anal. Chem.* **1987**, *59*, 44–49.

[3] Schmid, M.G.; Gübitz, G. Direct Enantiomer Separation of Underivatized Amino Acids by Capillary Zone Electrophoresis Based on Ligand Exchange, *Enantiomer* **1996**, *1*, 23–27.

[4] Végvári, A.; Schmid, M.G.; Kilár, F.; Gübitz, G. Chiral separation of α-amino acids by ligand-exchange capillary electrophoresis using N-(2-hydroxy-octyl)-L-4-hydroxyproline as a selector, *Electrophoresis* **1998**, *19*, 2109–2112.

[5] Chen, Z.; Lin, J.-M.; Uchiyama, K.; Hobo, T. Simultaneous Separation of Sixteen Positional and Optical Isomers of the Tryptophan Family by Ligand-Exchange Micellar Electrokinetic Chromatography, *Chromatographia* **1999**, *49*, 436–443.

[6] Chen, Z.; Lin, J.-M.; Uchiyama, K.; Hobo, T. Separation Behaviour of Amino Acid Enantiomers in Ligand Exchange Micellar Electrokinetic Chromatography, *J. Microcolumn Separations* **1999**, *11*, 534–540.

[7] Otsuka, K.; Terabe, S. Effects of methanol and urea on optical resolution of phenylthiohydantoin-DL-amino acids by micellar electrokinetic chromatography with sodium N-dodecanoyl-L-valinate, *Electrophoresis* **1990**, *11*, 982–984.

[8] Otsuka, K.; Kawahara, J.; Tatekawa, K.; Terabe, S. Chiral separations by micellar electrokinetic chromatography with sodium N-dodecanoyl-L-valinate, *J. Chromatogr.* **1991**, *559*, 209–214.

[9] Mazzeo, J.R.; Grover, E.R.; Swartz, M.E.; Petersen, J.S. Novel chiral surfactant for the separation of enantiomers by micellar electrokinetic chromatography, *J. Chromatogr. A* **1994**, *680*, 125–135.

[10] Wang, J.; Warner, I.M. Combined polymerized chiral micelle and γ-cyclodextrin for chiral separations in capillary electrophoresis, *J. Chromatogr. A* **1995**, *711*, 297–304.

[11] Shamsi, S.A.; Warner, I.M. Monomeric and polymeric chiral surfactants as pseudo-stationary phases for chiral separations, *Electrophoresis* **1997**, *18*, 853–872.

[12] Nishi, H.; Fukuyama, T.; Matsuo, M.; Terabe, S. Chiral Separation of Optical Isomeric Drugs Using Micellar Electrokinetic Chromatography and Bile Salts, *J. Microcolumn Separations* **1989**, *1*, 234–241.

[13] Amini, A.; Beijersten, I.; Pettersson, C.; Westerlund, D. Enantiomeric separation of local anaesthetic drugs by micellar electrokinetic capillary chromatography with taurodeoxycholate as chiral selector, *J. Chromatogr. A* **1996**, *737*, 301–313.

[14] Okafo, G.N.; Camilleri, P. Direct Chiral Resolution of Amino Acid Derivatives by Capillary Electrophoresis, *J. Microcolumn Separations* **1993**, *5*, 149–153.

[15] Otsuka, K.; Terabe, S. Enantiomeric resolution by micellar electrokinetic chromatography with chiral surfactants, *J. Chromatogr.* **1990**, *515*, 221–226.

[16] Mechref, Y.; Rassi, Z.E. Comparison of alkylglycoside surfactants in enantioseparation by capillary electrophoresis, *Electrophoresis* **1997**, *18*, 912–918.

[17] Ingelse, B.A.; Flieger, M.; Claessen, H.A.; Everaerts, F.M. Ergot alkaloids as chiral selectors in capillary electrophoresis determination of separation mechanism, *J. Chromatogr. A* **1996**, *755*, 251–259.

[18] Ingelse, B.A.; Reijenga, J.C.; Everaerts, F.M. Reversed determination of the formation constants of 1-allylterguride with mandelic acid optical isomers using capillary electrophoresis, *J. Chromatogr. A* **1997**, *772*, 179–184.

[19] Nair, J.; Armstrong, D.W.; Hinze, W.L. Characterization and Evaluation of d-(+)-Tubocurarine Chloride as a Chiral Selector for Capillary Electrophoretic Enantioseparations, *Anal. Chem.* **1998**, *70*, 1059–1065.

[20] Kuhn, R.; Erni, F.; Bereuter, T.; Häusler, J. Chiral Recognition and Enantiomeric Resolution Based on Host-Guest Complexation with Crown Ethers in Capillary Zone Electrophoresis, *Anal. Chem.* **1992**, *64*, 2815–2820.

[21] Kuhn, R.; Stoecklin, F.; Erni, F. Chiral Separations by Host-Guest Complexation with Cyclodextrin and Crown Ether in Capillary Zone Electrophoresis, *Chromatographia* **1992**, *33*, 32–36.

[22] Verleysen, K.; Vandijck, J.; Schelfaut, M.; Sandra, P. Enantiomeric Separations in Capillary Electrophoresis Using 18-Crown-6-tetracarboxylic Acid (18C6H4) as Buffer Additive, *J. High Resol. Chromatogr.* **1998**, *21*, 323–331.

[23] Verleysen, K.; Sandra, P. Enantiomeric Separation of Some Amino Acids and Derivatives by Capillary Electrophoresis with 18-Crown-6-Tetracarboxylic Acid as Chiral Selector, *J. Microcolumn Separations* **1999**, *11*, 37–43.

[24] Barker, G.E.; Russo, P.; Hartwick, R.A. Chiral Separation of Leucovorin with Bovine Serum Albumin Using Affinity Capillary Electrophoresis, *Anal. Chem.* **1992**, *64*, 3024–3028.

[25] Vespalec, R.; Sustacek, V.; Bocek, P. Prospects of dissolved albumin as a chiral selector in capillary zone electrophoresis, *J. Chromatogr.* **1993**, *638*, 255–261.

[26] Valtcheva, L.; Mohammad, J.; Pettersson, G.; Hjertén, S. Chiral separation of β-blockers by high-performance capillary electrophoresis based on non-immobilized cellulase as enantioselective protein, *J. Chromatogr.* **1993**, *638*, 263–267.

[27] Tanaka, Y.; Terabe, S.; Partial separation zone technique for the separation of enantiomers by affinity electrokinetic chromatography with proteins as chiral pseudo-stationary phases, *J. Chromatogr. A* **1995**, *694*, 277–284.

[28] Amini, A.; Pettersson, C.; Westerlund, D. Enantioresolution of disopyramide by capillary affinity electrokinetic chromatography with human α₁-acid glycoprotein

(AGP) as chiral selector applying a partial filling technique, *Electrophoresis* **1997**, *18*, 950–957.

[29] Haginaka, J.; Kanasugi, N. Separation of basic drug enantiomers by capillary zone electrophoresis using ovoglycoprotein as a chiral selector, *J. Chromatogr. A* **1997**, *782*, 281–288.

[30] Tanaka, Y.; Terabe, S. Studies on Enantioselectivities of Avidin, Avidin-Biotin Complex and Streptavidin by Affinity Capillary Electrophoresis, *Chromatographia* **1999**, *49*, 489–495.

[31] Tanaka, Y.; Matsubara, N.; Terabe, S. Separation of enantiomers by affinity electrokinetic chromatography using avidin, *Electrophoresis* **1994**, *15*, 848–853.

[32] Armstrong, D.W.; Rundlett, K.L.; Chen, J.-R. Evaluation of the Macrocyclic Antibiotic Vancomycin as a Chiral Selector for Capillary Electrophoresis, *Chirality* **1994**, *6*, 496–509.

[33] Armstrong, D.W.; Rundlett, K.; Reid III, G.L. Use of a Macrocyclic Antibiotic, Rifamycin B, Indirect Detection for the Resolution of Racemic Amino Alcohols by CE, *Anal. Chem.* **1994**, *66*, 1690–1695.

[34] Armstrong, D.W; Tang, Y.; Chen, S.; Zhou, Y.; Chen, J.-R.; Bagwill, C. Macrocyclic antibiotics as a new class of chiral selector for liquid chromatography, *Anal. Chem.* **1994**, *66*, 1473–1484.

[35] Carpenter, J.L.; Camilleri, P.; Dhanak, D.; Goodall, D. A study of the binding of vancomycin to dipeptides using capillary electrophoresis, *J. Chem. Soc. Chem. Commun.* **1992**, 804–806.

[36] Liu, J.; Volk, K.J.; Lee, M.S.; Pucci, M.; Handwerger, S. Binding Studies of Vancomycin to the Cytoplasmic Peptidoglycan Precursors by Affinity Capillary Electrophoresis, *Anal. Chem.* **1994**, *66*, 2412–2416.

[37] Vespalec, R.; Corstjens, H.; Billiet, H.A.H.; Frank, J.; Luyben, K.Ch.A.M. Enantiomeric Separation of Sulfur- and Selenium-Containing Amino Acids by Capillary Electrophoresis Using Vancomycin as a Chiral Selector, *Anal. Chem.* **1995**, *67*, 3223–3228.

[38] Rundlett, K.L.; Armstrong, D.W. Effect of Micelles and Mixed Micelles on Efficiency and Selectivity of Antibiotic-Based Capillary Electrophoretic Enantioseparations, *Anal. Chem.* **1995**, *67*, 2088–2095.

[39] Fanali, S.; Desiderio, C.; Aturki, Z. Enantiomeric resolution study by capillary electrophoresis Selection of the appropriate chiral selector, *J. Chromatogr. A* **1997**, *772*, 185–194.

[40] Fanali, S.; Desiderio, C.; Schulte, G.; Heitmeier, S.; Strickman, D.; Chankvetadze, B.; Blaschke, G. Chiral capillary electrophoresis-electrospray mass spectrometry coupling using vancomycin as chiral selector, *J. Chromatogr. A* **1998**, *800*, 69–76.

[41] Reilly, J.; Risley, D.S. The Separation of Enantiomers by Countercurrent CE Using the Macrocyclic Antibiotic A82846B, *LC-GC International* September, **1998**, 598–606.

[42] Armstrong, D.W.; Gasper, M.P.; Rundlett, K.L. Highly enantioselective capillary electrophoretic separations with dilute solutions of the macrocyclic antibiotic ristocetin A, *J. Chromatogr. A* **1995**, *689*, 285–304.

[43] Gasper, M.P.; Berthod, A.; Nair, U.B.; Armstrong, D.W. Comparison and Modeling Study of Vancomycin, Ristocetin A, Teicoplanin for CE Enantioseparations, *Anal. Chem.* **1996**, *68*, 2501–2514.

[44] Carotti, A.; Di Gioia, F.; Cellamare, S.; Fanali, S. Teicoplanin-Based Enantiomeric Separations in CZE Using a Partial Filling Technique, *J. High Resol. Chromatogr.* **1999**, *22*, 315–321.

[45] Wan, H.; Blomberg, L.G. Chiral separation of DL-peptides and enantioselective interactions between teicoplanin and D-peptides in capillary electrophoresis, *Electrophoresis* **1997**, *18*, 943–949.

[46] D' Hulst, A.; Verbeke, N. Chiral separation by capillary electrophoresis with oligosaccharides, *J. Chromatogr.* **1992**, *608*, 275–287.

[47] Chankvetadze, B.; Saito, M.; Yashima, E.; Okamoto, Y. Enantioseparation using selected polysaccharides as chiral buffer additives in capillary electrophoresis, *J. Chromatogr. A* **1997**, *773*, 331–338.

[48] Bowser, M.T.; Kranack, A.R.; Chen, D.D.Y. Analyte-additive interactions in non-aqueous capillary electrophoresis: a critical review, *TRAC* **1998**, *17*, 424–434.

[49] Wang, F.; Khaledi, M.G. Chiral separations by Nonaqueous Capillary Electrophoresis, *Anal. Chem.* **1996**, *68*, 3460–3467.

[50] Wang, F.; Khaledi, M.G. Nonaqueous capillary electrophoresis chiral separations with quaternary ammonium β-cyclodextrin, *J. Chromatogr. A* **1998**, *817*, 121–128.

[51] Vincent, J.B.; Vigh, G. Nonaqueous capillary electrophoretic separation of enantiomers using the single-isomer heptakis(2,3-diacetyl-6-sulfato)-β-cyclodextrin as chiral resolving agent, *J. Chromatogr. A* **1998**, *816*, 233–241.

[52] Bjrnsdottir, I.; Hansen, S.H.; Terabe, S. Chiral separation in non-aqueous media by capillary electrophoresis using the ion-pair principle, *J. Chromatogr. A* **1996**, *745*, 37–44.

[53] Karbaum, A.; Jira, T. Chiral separation of unmodified amino acids with non-aqueous capillary electrophoresis based on the ligand-exchange principle, *J. Chromatogr. A* **2000**, *874*, 285–292.

[54] Mardones, C.; Rios, A.; Valcárcel, M.; Cicciarelli, R. Enantiomeric separation of D- and L-carnitine by integrating on-line derivatisation with capillary electrophoresis, *J. Chromatogr. A* **1999**, *849*, 609–616.

[55] Thorsén, G.; Engström, A.; Josefsson, B. Enantiomeric determination of amino compounds with high sensitivity using the chiral reagents (+)- and (–)-1-(9-anthryl)-2-propyl chloroformate, *J. Chromatogr. A* **1997**, *786*, 347–354.

[56] Liu, Y.-M.; Schneider, M.; Sticha, C.M.; Toyooka, T.; Sweedler, J.V. Separation of amino acid and peptide stereoisomers by nonionic micelle-mediated capillary electrophoresis after chiral derivatisation, *J. Chromatogr. A* **1998**, *800*, 345–354.

[57] Jorgenson, J.W.; Lukacs, K.D. High-resolution separations, based upon electrophoresis and electroosmosis, *J. Chromatogr.* **1981**, *218*, 209–216.

[58] Neue, U.D. HPLC Columns Theory, Technology, Practice, *Wiley-VCH*, New York, **1997**.

[59] Knox, J.H.; Grant, I.H. Electrochromatography in Packed Tubes Using 1.5 to 50 μm Silica and ODS Bonded Silica Gels, *Chromatographia* **1991**, *32*, 317–328.

[60] Adam, Th.; Lüdtke, S.; Unger, K.K. Packing and Stationary Phase Design for Capillary Electroendosmotic Chromatography (CEC), *Chromatographia* **1999**, *49*, 49–55.

[61] Wen, E.; Asiaie, R.; Horváth, Cs. Dynamics of capillary electrochromatography II. Comparison of column efficiency parameters in microscale high-performance liquid chromatography and capillary electrochromatography, *J. Chromatogr. A* **1999**, *855*, 349–366.

[62] Li, S.; Lloyd, D.K. Direct Chiral Separations by Capillary Electrophoresis Using Capillaries Packed with α₁-Acid Glycoprotein Chiral Stationary Phase, *Anal. Chem.* **1993**, *65*, 3684–3690.

[63] Carter-Finch, A.S.; Smith, N.W. Enantiomeric separations by capillary electrochromatography using a macrocyclic antibiotic chiral stationary phase, *J. Chromatogr. A* **1999**, *848*, 375–385.

[64] Wolf, C.; Spence, P.L.; Pirkle, W.H.; Derrico, E.M.; Cavender, D.M.; Rozing, G.P. Enantioseparations by electrochromatography with packed capillaries, *J. Chromatogr. A* **1997**, *782*, 175–179.

[65] Lämmerhofer, M.; Lindner, W. High-efficiency chiral separations of N-derivatized amino acids by packed-capillary electrochromatography with a quinine-based chiral anion-exchange type stationary phase, *J. Chromatogr. A* **1998**, *829*, 115–125.

[66] Guttman, A.; Paulus, A.; Cohen, A.S.; Grinberg, N.; Karger, B.L. Use of complexing agents for selective separation in high-performance capillary electrophoresis. Chiral resolution via cyclodextrins incorporated within polyacrylamide gel columns, *J. Chromatogr.* **1988**, *448*, 41–53.

[67] Koide, T.; Ueno, K. Enantiomeric Separations of Acidic and Neutral Compounds by Capillary Electrochromatography with β-Cyclodextrin-Bonded Positively Charged Polyacrylamide Gels, *J. High Resol. Chromatogr.* **2000**, *23*, 59–66.

[68] Nilsson, S.; Schweitz, L.; Petersson, M. Three approaches to enantiomer separation of β-adrenergic antagonists by capillary electrochromatography, *Electrophoresis* **1997**, *18*, 884–890.

[69] Mayer, S.; Schurig, V. Enantiomer Separation by Electrochromatography on Capillaries Coated With Chiralsil-Dex. *J. High Resol. Chromatogr.* **1992**, *15*, 129–131.

[70] Mayer, S.; Schurig, V. Enantiomer separation by electrochromatography in open tubular columns coated with Chirasil-Dex. *J. Liq. Chromatogr.* **1993**, *16*, 915–931.

Chromatographia Supplement Vol. 54, 2001, Index

A82846B , **54**, S-85
Absolute configuration, **54**, S-8
Acidic analytes, **54**, S-48, S-66, S-85
Acid-base equilibria, **54**, S-39
Accuracy, **54**, S-57
Adrenaline, **54**, S-67
Adrenoreceptor antagonists, **54**, S-66
Alanine, **54**, S-13, S-80, S-88
Alkylglycoside surfactants, **54**, S-80
Alkyl pyridines, **54**, S-16, S-17, S-18
Allylterguride, **54**, S-81
Alpha1-acid glycoprotein, **54**, S-82, S-83, S-90
Alprenolol, **54**, S-48
Amino acids, **54**, S-8, S-78, S-79, S-80, S-81, S-84, S-87, S-88
Aminoacridone, **54**, S-73
Aminobutyric acid, **54**, S-51, S-80
Amphetamaines, **54**, S-55, S-72, S-73
Amylose, **54**, S-86
Anaesthetics, **54**, S-68, S-80
Analgesic, **54**, S-72
Angina pectoris, **54**, S-9
Anthrylpropylchloroformate, **54**, S-88
Anticholinergics, **54**, S-68
Anti-clockwise rotation, **54**, S-7
Antidepressants, **54**, S-68
Antihistamines, **54**, S-62, S-68
Antimalarials, **54**, S-68
Apparent electrophoretic mobility, **54**, S-26, S-27, S-28, S-29, S-31
APTS, **54**, S-45
Area %, **54**, S-57
Array screening, **54**, S-48
Arrhythmia, **54**, S-10
Arylpropionic acids, **54**, S-9, S-84
Aspartic acid, **54**, S-48, S-70, S-73, S-80, S-81, S-82
Aspartame, **54**, S-78
Asymmetric centre, **54**, S-8, S-10, S-81
Asymmetric synthesis, **54**, S-8
Atenolol, **54**, S-32, S-33, S-34, S-36, S-44, S-47, S-48, S-67
Atropa belladonna, **54**, S-73
Atropine, **54**, S-73
Atropisomers, **54**, S-74
Avidin, **54**, S-83

Baclofen, **54**, S-80
Band broadening, **54**, S-22, S-26, S-53, S-89
Barbituates, **54**, S-13, S-64, S-69, S-71, S-90
Baseline resolution, **54**, S-26, S-56, S-86, S-87

Basic analytes, **54**, S-48, S-56, S-66, S-70, S-71, S-85, S-87, S-90
Beer-Lambert law, **54**, S-22
Benzodiazepines, **54**, S-61
Benzoin, **54**, S-69, S-79
Beta-blockers, **54**, S-8, S-9, S-25, S-27, S-32, S-33, S-44, S-66, S-67, S-86
Bile acids, **54**, S-79
1, S-1'-bi-naphthol, **54**, S-13, S-68, S-75, S-79
Binding sites, **54**, S-83, S-85
Biogenic amines, **54**, S-63
Biotin, **54**, S-83
Borate complexation, **54**, S-69, S-70
Bovine Serum Albumin (BSA), **54**, S-82, S-83
Boyle's law, **54**, S-24
Bronchodilators, **54**, S-68, S-86
Buffer temperature, **54**, S-63
Buffering capacity, **54**, S-50
Bulk drug, **54**, S-10,
Bupivacaine, **54**, S-10, S-47, S-71, S-75, S-80
Butansultone, **54**, S-67, S-68

Cahn-Ingold-Prelog rules, **54**, S-8
Camphor sulphonate, **54**, S-70, S-71, S-87
Capillary wall, **54**, S-18
Carboxylic acids, **54**, S-43, S-83, S-84
Carnitine, **54**, S-88
Carprofen, **54**, S-69
Cathonine, **54**, S-73
CEC, **54**, S-11, S-88, S-89, S-90, S-91
Celluloses, **54**, S-86
Central composite design, **54**, S-55
Chiral pool, **54**, S-8
Chiral selector, **54**, S-10, S-11
Chiral stationary phase, **54**, S-10
Chiral switch, **54**, S-10
Chloramphenicol, **54**, S-12
Chlorthalidone, **54**, S-69
Clenbuterol, **54**, S-50, S-53, S-54, S-55, S-57, S-69
Clockwise rotation, **54**, S-7
Cocaine, **54**, S-72, S-73
Combinatorial synthesis, **54**, S-48
Complexation stoichiometries, **54**, S-61
Computer modelling, **54**, S-26, S-30, S-40
Computer simulation, **54**, S-57
Computer software, **54**, S-57
Configuration, **54**, S-8, S-87
Corrected areas, **54**, S-23, S-57
Cross validation, **54**, S-57
Curare, **54**, S-81

Cyclic hexapeptides, **54**, S-48
Cyclodextrin glucosyl transferase, **54**, S-60
Cyclodrine, **54**, S-71

Dansylated amino acids, **54**, S-11, S-12, S-13, S-24, S-65, S-66, S-67, S-69, S-84
Data collection, **54**, S-21
Deadly nightshade, **54**, S-73
Debye-Hückel parameter, **54**, S-19
Degree of dissociation, **54**, S-51
Degree of substitution, **54**, S-67, S-68
Derivatisation procedures, **54**, S-68
Desmethylvenlafaxine, **54**, S-71, S-72
Desionselective mechanism, **54**, S-64
Detector, **54**, S-21, S-22
Dexrotatory, **54**, S-8
Dextran ladder, **54**, S-45
Diastereoisomers, **54**, S-10, S-18, S-37, S-73, S-88
Diazepam, **54**, S-60
Dichloroprop, **54**, S-10, S-73, S-74
Differential inclusion, **54**, S-61
Diffusion coefficient, **54**, S-26, S-42, S-89
Digitonin, **54**, S-80
Dihydroxyphenylalanine (DOPA), **54**, S-79
Diltiazem, **54**, S-79, S-86
Dimethindene, **54**, S-62
Dimethyl pyridine, **54**, S-17
Dinitrophenylhydrazine, **54**, S-48
Diols, **54**, S-70
Dioxypromethazine, **54**, S-40
Disopyramide, **54**, S-67, S-71
Dispersive transport, **54**, S-15
Disulphonated benzene, **54**, S-43
Double layer thickness, **54**, S-19
Drug substance, **54**, S-42
Dual cyclodextrin systems, **54**, S-55, S-56
Duloxetine, **54**, S-67
Duoselective mechanism, **54**, S-64

Efficiency, **54**, S-16, S-21, S-32, S-34, S-43, S-50, S-52, S-82, S-84, S-89
Electrochemical detection, **54**, S-22
Electrokinetic injection, **54**, S-22
Electrolysis, **54**, S-50
Electroosmosis, **54**, S-16
Electrophoresis, **54**, S-16
Electrophoretic dispersion, **54**, S-42, S-50, S-51, S-69
Electrophoretic mobility difference, **54**, S-25, S-26, S-28, S-31, S-32, S-69, S-84, S-85

Elution order, **54**, S-64
Enantiomeric metabolites, **54**, S-9
Enantioselectivity, **54**, S-59, S-62, S-63, S-64, S-65, S-66, S-67, S-68, S-70, S-75, S-90
Enthalpy, **54**, S-63
Entropy, **54**, S-15, S-53, S-63
Ephedrine, **54**, S-12, S-48, S-67, S-73, S-79, S-81
Epinastine, **54**, S-83
Epinephrine, **54**, S-12, S-48, S-55, S-57
Equilibrium complex, **54**, S-26
Equilibrium reaction, **54**, S-25, S-27
Ergot alkaloids, **54**, S-73, S-80
Ethyl pyridines, **54**, S-17
Etodolac, **54**, S-84, S-85
Experimental design, **54**, S-54, S-55, S-56, S-73

Factorial design, **54**, S-57
Fast kinetics, **54**, S-59
Fenoprofen, **54**, S-48, S-64, S-68, S-71, S-87
Fischer projection, **54**, S-8
Flow chart, **54**, S-55, S-56
Flow profile, **54**, S-20, S-90
Fluoroenylmethylchloroformate (FMOC), **54**, S-88
Fluorescamine, **54**, S-11
Fluorescence detection, **54**, S-12, S-22, S-45, S-73
Fluoromandelic acid, **54**, S-83
Fluorophenylalanine, **54**, S-79
Flurbiprofen, **54**, S-69, S-83, S-84, S-86, S-91
Fluparoxan, **54**, S-57
Fronting, **54**, S-50, S-51, S-64
Food and drug administration (FDA), **54**, S-10
Formulated product, **54**, S-10
Fractional crystallisation, **54**, S-10
Fractional designs, **54**, S-55
Fucose, **54**, S-73

Galactose, **54**, S-73
GC, **54**, S-10, S-11
Gibbs free energy, **54**, S-53, S-59, S-64
Glucose oligomers, **54**, S-45
Glutamic acid, **54**, S-48, S-51, S-69, S-70, S-80, S-81
Glyceraldehyde, **54**, S-8, S-21
Glycine, **54**, S-88
Good's buffers, **54**, S-20
Guest molecules, **54**, S-60
Gulonic acid, **54**, S-72

Herbicides, **54**, S-73, S-74
Hexobarbital, **54**, S-69, S-75
Hindered rotation, **54**, S-74, S-75
Histidine, **54**, S-12, S-78, S-80, S-81
Histamine, **54**, S-69

Homatropine, **54**, S-48
HPLC, **54**, S-10, S-11, S-12, S-16, S-59, S-64, S-75, S-82, S-90
Human Serum Albumin (HSA), **54**, S-82
Hydantoins, **54**, S-64
Hydrobenzoin, **54**, S-70
Hydrocortisone, **54**, S-61
Hydrodynamic injection, **54**, S-22
Hydrophobic interactions, **54**, S-61, S-87
Hydrostatic injection, **54**, S-22
Hydroxyisobutyric acid, **54**, S-51
Hydroxyphenylacetic acid, **54**, S-51
Hydroxyproline, **54**, S-78, S-79
Hyoscyamine, **54**, S-73
Hyoscyamus albus, **54**, S-73

Ibuprofen, **54**, S-9, S-48, S-63, S-64, S-71, S-84, S-85, S-86, S-91
Imidazole, **54**, S-69
Indoprofen, **54**, S-68, S-84, S-87
Injection volume, **54**, S-54
Intermediate precision, **54**, S-57
Internal standard, **54**, S-57
Inverse detection, **54**, S-86
Ionselective mechanism, **54**, S-64
Ionic equilibria, **54**, S-18
Ion pair reagents, **54**, S-71, S-87
Ionic strength, **54**, S-18, S-42, S-49, S-50
Isoelectric point, **54**, S-83, S-84, S-85
Isoleucine, **54**, S-87
Isoproterenol, **54**, S-12

Joule heating, **54**, S-21, S-25, S-53, S-54

Laevorotatory, **54**, S-8
Laser induced fluorescence, **54**, S-45, S-73, S-78, S-88
Leucine, **54**, S-13, S-48, S-65, S-80, S-90
Leucovorin, **54**, S-82
Limit of detection , **54**, S-57, S-84
Limit of quantification, **54**, S-57
Limiting mobilities, **54**, S-18, S-26, S-27, S-30, S-31, S-36, S-37, S-38, S-44
Limonene, **54**, S-8
Linear velocity, **54**, S-89, S-90, S-91
Linearity, **54**, S-57
Liquid cooling, **54**, S-21
Local anaesthetic, **54**, S-10, S-47, S-57
Log P, **54**, S-61
LY248686, **54**, S-50, S-51
Lysergic acid, **54**, S-80

MALDI-MS, **54**, S-68
Maltodextrins, **54**, S-86
Mandelic acid, **54**, S-9, S-62, S-68, S-80, S-85
Mass transfer, **54**, S-89, S-90, S-91
Maximum mobility difference, **54**, S-33, S-48
MEKC, **54**, S-13, S-25, S-79
Menthoxyacetic acid, **54**, S-70

Mepivacaine, **54**, S-47, S-80
Metabolic inversion, **54**, S-9
Metabolites, **54**, S-72, S-84
Metaproterenol, **54**, S-48
Method validation, **54**, S-56, S-57, S-75
Methoxyphenamine, **54**, S-48
Methyl pyridine, **54**, S-17, S-49
Methionine, **54**, S-13, S-65, S-80, S-84
Metoprolol, **54**, S-32, S-33, S-34, S-36, S-42, S-86, S-87
Mianserin, **54**, S-86
Microcalorimetry, **54**, S-63, S-70
Migration order, **54**, S-12, S-38, S-45, S-65, S-75, S-78, S-79, S-80, S-87, S-88
Mirror images, **54**, S-7
Minor enantiomer, **54**, S-42
Mixed selector systems, **54**, S-69
MK-0677, **54**, S-70
Mobility matching, **54**, S-51, S-52
Molar extinction coefficient, **54**, S-57
Molecular diffusion, **54**, S-26, S-40
Molecular imprinting, **54**, S-91
Mole fraction, **54**, S-26
Monosaccharides, **54**, S-73
MS detection, **54**, S-22, S-50, S-72, S-75, S-84
Multiple regression, **54**, S-55
Multivariate optimisation, **54**, S-53, S-54, S-55

Naproxen, **54**, S-48, S-65, S-84, S-90
Naphthalene-2, S-3-dicarboxaldehyde, **54**, S-13
Naphthylethylamine, **54**, S-89
Neutral marker, **54**, S-20
^{13}C NMR, **54**, S-59, S-62, S-75
^{19}F NMR, **54**, S-25
^{1}H NMR, **54**, S-25, S-61, S-62, S-75
N-oxides, **54**, S-74
Non-steroidal-anti-inflammatory-drugs (NSAIDs), **54**, S-9, S-68, S-71, S-84, S-85, S-86
Noradrenaline, **54**, S-67, S-81
Norcardia mediterranei, **54**, S-85
Norephedrine, **54**, S-12, S-72, S-73, S-81
Norepinephrine, **54**, S-12,
Norleucine, **54**, S-13, S-79, S-80
Norpseudoephedrine, **54**, S-72, S-73
Norvaline, **54**, S-13, S-65, S-70, S-79, S-80
Nuclear Overhauser experiments, **54**, S-62

Offord's parameter, **54**, S-16, S-17
Ohm plot, **54**, S-22
Oligoalanines, **54**, S-17, S-18
Oligoglycines, **54**, S-17, S-18
Oligosaccharides, **54**, S-86
Ophthalmic solutions, **54**, S-73
Optical activity, **54**, S-7
Optical purity, **54**, S-49, S-64
Optimisation schemes, **54**, S-53

Optimum chiral selector concentration, **54**, S-25, S-29, S-30, S-31, S-32, S-33, S-34, S-35, S-38, S-48, S-63

Overloading, **54**, S-42, S-49, S-64

Ovoglycoprotein, **54**, S-83

Ovomucoid, **54**, S-83

Oxprenolol, **54**, S-32, S-33, S-34, S-36, S-44, S-86

Partial filling method, **54**, S-83, S-84, S-85

Partial separation zone, **54**, S-82

Pasteur, **54**, S-7

Peak dispersion, **54**, S-43, S-46, S-69

Peak tailing, **54**, S-42, S-50, S-51, S-56, S-64, S-79, S-91

Pentobarbital, **54**, S-25, S-70, S-71

Peptides, **54**, S-39, S-73, S-80, S-85, S-88

pH, **54**, S-18, S-20, S-40, S-49, S-63, S-64, S-68, S-69, S-79, S-85

Pharmacodynamic, **54**, S-9

Pharmacokinetic, **54**, S-9

Phenoxy acid herbicides, **54**, S-73

Phenylacetic acid, **54**, S-51

Phenylalanine, **54**, S-13, S-48, S-69, S-78, S-80, S-81, S-87

Phenylglycine, **54**, S-78

Phenylthiohydantoin derivatives, **54**, S-79, S-80, S-84

Phenyllactic acid, **54**, S-51, S-68

Phenyramidol, **54**, S-51, S-52

Picumeterol, **54**, S-50, S-54

pK_a, **54**, S-18, S-40, S-49, S-64

Plackett-Burman experimental design, **54**, S-55

Plane polarised light, **54**, S-7

Plate count, **54**, S-43, S-44, S-52

Poisseuille equation, **54**, S-22

Polychlorinated biphenyls, **54**, S-75

Polymeric surfactants, **54**, S-79

Polysaccharides, **54**, S-86

Power supply, **54**, S-21

Practolol, **54**, S-47

Precision, **54**, S-57

Prilocaine, **54**, S-47, S-80

Product enantioselectivity, **54**, S-9

Proline, **54**, S-78, S-87

Propanolol, **54**, S-8, S-9, S-12, S-24, S-25, S-27, S-28, S-32, S-33, S-35, S-36

Pseudoephedrine, **54**, S-57, S-67

Pyrethroids, **54**, S-10

Quantification, **54**, S-26, S-42

Quinine, **54**, S-71, S-90

Racemic acid, **54**, S-7

Racemic mixture, **54**, S-7, S-8, S-48

Racemisation, **54**, S-9, S-73

Raman spectroscopy, **54**, S-59

Reflectional symmetry, **54**, S-7

Related substances, **54**, S-42, S-71

Repeatability, **54**, S-57

Reproducible analysis conditions, **54**, S-20

Resmethrin, **54**, S-10

Response surfaces, **54**, S-55

Ribose, **54**, S-73

Rifamycin B, **54**, S-85, S-86

Ristocetin A, **54**, S-85

Robustness, **54**, S-56, S-57, S-82, S-90, S-91

Ropivacaine, **54**, S-47, S-57

Salbutamol, **54**, S-47, S-69, S-75, S-87

Sample matrix, **54**, S-42

Sample overloading, **54**, S-26, S-51

Sample concentration, **54**, S-54

Sample loadings, **54**, S-43

Scouting experiments, **54**, S-56

Screening experiments, **54**, S-48, S-49, S-54, S-55, S-56, S-57, S-66,

Scopolia japonica, **54**, S-73

Secobarbital, **54**, S-71

Selectivity, **54**, S-16, S-44

Selective synthesis, **54**, S-7

Selenomethionine, **54**, S-84

Separation, **54**, S-15

Separation efficiency, **54**, S-53

Separation methods, **54**, S-10

Separative transport, **54**, S-15, S-16

Serine, **54**, S-80

Silanol groups, **54**, S-18

Single isomer chiral selectors, **54**, S-68

SFC, **54**, S-10, S-11

Sodium *d*-camphor-10-sulphonate, **54**, S-13

Sodium cholate, **54**, S-80

Sodium deoxycholate, **54**, S-80

Sodium *l*-menthoxyacetate, **54**, S-13

Sodium *N*-dodecanoyl-L-valinate, **54**, S-13, S-79

Sodium *N*-dodecoxycarbonylvalinate, **54**, S-79

Sodium *N*-undecylenyl-L-valinate, **54**, S-79

Sodium taurodeoxycholate, **54**, S-80

Solid state NMR, **54**, S-59

Solvent polarity, **54**, S-62

Sotalol, **54**, S-51, S-52

Spironolactone, **54**, S-60

Specificity, **54**, S-57

Stationary phase, **54**, S-89, S-90, S-91

Stereogenic centre, **54**, S-74

Stereoselective binding, **54**, S-48

Stereoselective synthesis, **54**, S-10

Stereoselective reduction, **54**, S-72

Streptomyces orientalis, **54**, S-83

Substrate enantioselectivity, **54**, S-9

Sugars, **54**, S-8

Sulindac, **54**, S-74

Sulphonium ions, **54**, S-75

Supercritical fluid extraction, **54**, S-73

Symmetrical peaks, **54**, S-52

Symmetry, **54**, S-7

Systematic variation, **54**, S-54

Tartaric acid, **54**, S-7, S-10, S-70

Teicoplanin, **54**, S-85, S-90

Temperature control, **54**, S-12,

Terbutaline, **54**, S-12, S-24, S-48, S-62, S-69, S-75, S-86, S-87

Tetrapeptide enantiomers, **54**, S-39

Thalidomide, **54**, S-9, S-69

Theoretical plates, **54**, S-32, S-43

Thermal gradients, **54**, S-11, S-21, S-44, S-53

Thermodynamics, **54**, S-63

Thiopental, **54**, S-25, S-70

Thioridazine, **54**, S-12, S-86

Threonine, **54**, S-70, S-80

Tioconazole, **54**, S-36, S-48, S-62

TLC, **54**, S-10, S-59

Tocainide, **54**, S-67

Tramadol, **54**, S-72

Transient complexes, **54**, S-26, S-36

Transport processes, **54**, S-16

2,2,2-trifluoro-1-(9-anthryl)-ethanol, **54**, S-25

Trimetoquinol, **54**, S-80

Trimipramine, **54**, S-86

Trögers base, **54**, S-74

Tryptophan, **54**, S-24, S-70, S-79, S-80, S-81, S-87, S-90

Tubocurarine, **54**, S-81

Type I enantiomers, **54**, S-40, S-64

Type II enantiomers, **54**, S-40, S-64

Type III enantiomers, **54**, S-40, S-64

Tyrosine, **54**, S-67, S-78, S-81

Univariate optimisation, **54**, S-53, S-54, S-55

Validation, **54**, S-56, S-57, S-88

Valine, **54**, S-13, S-70, S-79, S-80, S-84, S-86

Vancomycin, **54**, S-83, S-84, S-85, S-86

Van der Waals radius, **54**, S-18

Van't Hoff isochore, **54**, S-63

Venlafaxine, **54**, S-71, S-72

Verapamil, **54**, S-83

Warfarin, **54**, S-69, S-79, S-80, S-86

X-ray diffraction, **54**, S-59, S-61

Z-shaped flow cell, **54**, S-22

Zeta potential (ξ), **54**, S-20, S-88, S-89

Zwitterionic buffers, **54**, S-50

Zwitterionic species, **54**, S-68, S-69